The Forgotten Botanist

Sara Plummer Lemmon's Life of Science and Art

WYNNE BROWN

University of Nebraska Press | LINCOLN

Library of Congress Cataloging-in-Publication Data
Names: Brown, Wynne L., author.
Title: The forgotten botanist: Sara Plummer
Lemmon's life of science and art / Wynne Brown.
Description: Lincoln: University of Nebraska Press,
[2021] | Includes bibliographical references.
Identifiers: LCCN 2021008843
ISBN 9781496222817 (paperback)
ISBN 9781496229465 (epub)
ISBN 9781496229472 (pdf)
Subjects: LCSH: Lemmon, Sara Allen Plummer,
1836–1923. | Women botanists—United States—
Biography. | Botanical illustration.
Classification: LCC QK31.L455 B76 2021 |
DDC 580.92 [B]—dc23
LC record available at
https://lccn.loc.gov/2021008843

Set in Arno Pro by Mikala R. Kolander.

To all the hidden women of science and
art whose stories have not yet been told

Contents

Illustrations

Preface

↝ PEOPLE OFTEN ASK ME, "Why do you care so much about Sara Lemmon?"

During the seven years' work on this project, I've wondered the same thing. Here's one answer. I've always been fascinated by that shaded area within a Venn diagram, the place where seemingly disparate disciplines swirl and mingle—like the pouring of cream into iced coffee. Sara's ferocious curiosity led her to some of those blends: science versus art, wilderness versus safety, the comforts of tradition versus the kick of questioning the status quo. Her curiosity, tenacity, and grit carried her farther than the Arizona Territory—where a mountain would be named for her—and deep into the territory of the unexpected.

I also admire her courage. In the twenty-first century, many of us mull reinventing ourselves. But moving from one coast to the other in the nineteenth century meant not seeing—or even hearing the voices of—your loved ones for five or six years. That takes more than reinvention.

It takes guts.

Sara was also immensely talented. Her few surviving watercolors are more than scientifically accurate renditions of the species: They're truly exquisite.

How could I *not* care about someone so intriguing?

In the early 2000s, while researching other remarkable pioneer women, I learned that the University of California and Jepson Herbaria Archives had six linear feet of field notes, correspondence, photographs, and art-work by Sara and John Lemmon. As someone who's worked as a journalist and a scientific researcher and illustrator, I itched to see that material.

Several years later, when I finally saw the stacks of folders containing Sara's handwritten letters and a tiny sample of her watercolors, I immediately knew they were a treasure—and that I lacked the historical training to know what to do with it.

When in doubt, capture.

So I made three trips to Berkeley from Arizona and photographed all twelve hundred pages of her letters. Over the next three years, I read every one of them.

It's no surprise that Sara Lemmon was a member of the California Press Women: She was a born reporter. Her voice is strong and consistent and lush with narrative details—whether she was describing the "placid" Mojave Indians, the scurrying of the overworked and underpaid copy girls at the *New York Independent*, the cool thickness of Santa Barbara's adobe walls, the sweat dripping off men's bare chests in Tombstone's mines, the vital need for conserving California's forests, the stars peeping through the silver fir branches in Kings Canyon, or how to cook lunch on a scrap of sheet metal in Baja California.

John Gill Lemmon, Sara's husband, is still known to his descendants as JG, so I've used that name as well. He too was an engaging and prolific writer. When Sara's letters were missing, I've used some of his writing to fill the gaps, while remembering he could be given to not letting the facts get in the way of a better story.

This book includes many excerpts of Sara's letters. To avoid inflicting my voice on hers, I've made very few changes, other than punctuation and updated spelling for clarity and easier reading. A warning: Some of her descriptions and attitudes are disturbing to our modern sensibilities. Sometimes she comes across as colonialist and extractive at best, and even downright racist at worst. I've had to remind myself to read her words while remembering the context—and the fears, however unfounded—of the times. I hope readers will do the same.

Sara Allen Plummer Lemmon's story is a window into one woman's early West. Losing both that narrative and her artistic legacy would be a tragedy.

How could I *not* care about her?

The Forgotten Botanist

Prologue

⇢ ICY BITS OF SLEET peppered Sara's face and caught in the curls of her auburn hair. She tried turning her shoulder into the fierce December wind that buffeted her coat against her legs, to no avail—and still she waved, waved until her arm ached, and kept waving, long after the figures of her sister and her father were too tiny to discern among the crowd of well-wishers on the harbor dock.

The wooden side-wheel steamer SS *Alaska* churned slowly through the steel-gray waters off Manhattan, away from Brooklyn where Sara had graduated from Greenleaf Female Institute with a teaching certificate, away from the Cooper Union Institute where she'd worked hard to earn degrees in physics and chemistry, away from the New England coast where she'd spent her childhood, away from Bellevue Hospital where she'd nursed so many Civil War soldiers, away from a myriad of friends, away from all the students who'd learned to paint thanks to the art lessons she provided.

Away from her loving, supportive family: her three brothers but especially her father and Mattie, her only sister. Away from a man who worshiped her and called her "my beloved." Away from everything and everyone she'd ever known—leaving it all behind.

Sara remained on the deck, shivering in the wind, long after every other passenger had retreated below.

After years of juggling her teaching, night classes, and volunteer nursing work, she thought she knew about being tired. But right now she was exhausted beyond any fatigue she'd ever known. She was tired of being sick, tired of packing, tired of saying goodbye, tired of trying to convince loved ones her very survival depended on starting a whole new life—a life that had to be an entire continent away. A life in exile from the contented busy life she had relished in New York City.

Mattie was two years younger than Sara and had married eight years previously. By now she and her husband had two daughters. Several months after arriving in California, Sara, whose nickname was "Sadie," wrote one of her young nieces:

> Dear Little Goody Two Shoes,
>
> Your Aunt Sadie wants to come home after she gets well, but it won't do in a great many weeks yet because she might get sick again and die. But after I get well, some time we will all get together, and you and your sister will have a tea party, and then I can come to it, and won't we have a good time!

There's no doubt that Sara's fragile health was the driving force for her new start in California. In 1863 she'd barely survived measles, and every winter she was flattened by colds, chronically inflamed sinuses, and bronchitis. Then in March 1869 she nearly died of pneumonia. She faced reality: One more northeastern winter would probably kill her. Even in a milder climate, her lungs would remain vulnerable for the rest of her life.

Standing on that windswept deck, Sara was frightened—and yet excited. Another decade was receding along with the New York coastline, a decade that marked the end of the Pony Express and the beginning of paper money in the United States, a decade in which E. Remington and Sons developed the first typewriter and Alfred Nobel patented his new mixture known as dynamite.

The decade had also brought the end of slavery and the untimely death of President Abraham Lincoln. Within the past year, Ulysses S. Grant had been sworn in as the eighteenth president of the United States, and Elizabeth Cady Stanton was the first woman to testify before Congress.

In a few weeks it would be 1870, and Sara had just turned thirty-three years old. Too old to marry, her brother Osgood had told her (she would prove him wrong), but surely she was still young enough to begin anew, wasn't she? What adventures, what joys, what tragedies would the next year, the next decade, the rest of her life bring?

Decades later Sara was described as determined to the point of being "firm and aggressive." Yet had she known the challenges of moving cross-country, had she foreseen the financial struggles and multiple near brushes with death—not to mention the loneliness and homesickness—she might have reconsidered her strategy.

But then she'd have missed out on adventure, exploration, and recognition for her work, although much less than she deserved. And most of all, she'd have missed out on a remarkable partnership—and love.

A year later, on December 27, 1870, she wrote Mattie from Santa Barbara:

> What a homesick Christmas I had this year. Not quite as forlorn as last year on board the steamer Alaska in the Caribbean Sea, watching the flying fish and porpoises. Wonder where I shall be next year—I dare not think of it. I hope the next year will be freighted with better things than the past has been. This old year seems like some horrid dream.
>
> But I have good courage and hope, feel very well, and have many friends here.

The next year would indeed bring better things, and the young woman who couldn't stand to be idle would find herself more than busy.

The new American West would be the beneficiary in many ways.

Fig. 1. Sara Allen Plummer in her late twenties, a decade before she left New York City for California. Photo by author. Original at the UC and Jepson Herbaria Archives, University of California, Berkeley.

1

"Now I Am at the Jumping Off Place"

San Francisco, 1870

→ FROM ALL ACCOUNTS, HUNDREDS of reporters, workers, and onlookers attended the driving of the Golden Spike at Promontory Summit in the Utah Territory on Monday, May 10, 1869. It was, after all, the hammering in of the last connection in the first transcontinental railroad that joined the Atlantic and Pacific oceans.

Although she had been a devoted newspaper reader since childhood, Sara Plummer probably missed the Golden Spike article that particular Monday. She'd nearly died of pneumonia two months earlier and was just returning to Manhattan after convalescing in Florida. Now she headed home to the house on East Twelfth Street where she boarded with several other women. It was a happy place to live and a lively gathering spot for noted lecturers and poets, including Phoebe and Alice Carey and Mrs. Fanny Gage. Sara's plan was to rest a few days, then try to return to her job at Ward School No. 14 teaching calisthenics (known to many of us as gym). As her energy increased, she'd resume teaching private art lessons as well. But by the end of summer and into fall, her health deteriorated yet again. In November, according to her sister, she had yet another attack of "that dreaded inflammation of the lungs."

For Sara, the new transcontinental railway wasn't just a national story—it would become personal. She needed a warmer, drier climate. She needed to move to California.

How and when should she make the transcontinental move?

Eventually Sara decided that passenger travel by railroad all the way across the country was still too iffy. Six months after the pounding in of the Golden Spike, she was among the last few hardy souls who journeyed to San Francisco by boat via Panama.

Several shipping lines ran regular routes from New York to Aspinwall (now known as Colón) on the east side of Panama, and in the twenty-one years from 1848 to 1869, 375,000 other California-bound travelers had chosen that route.

Sara traveled by steamer from New York down the East Coast, eventually reaching Aspinwall. The Panama Canal wouldn't open for another forty-five years, so she spent a day crossing the forty-seven-mile isthmus on a "smooth, well-kept track" through the malaria- and yellow-fever-ridden swamp and jungle. She told Mattie that it was like riding through a conservatory lined with every kind of tropical fruit tree, and "in many places, the banks are covered with sensitive plants [*Mimosa pudica*], in full bloom. It has fine green leaves like the locust, only not half as large, and at the touch, the pairs of leaves close. The flower is round, and pink, fine like the head of a mouse ear, the color of your Spirea."

After arriving in Panama City on the west coast, she was able to arrange passage on another steamer, the *Colorado*, that churned its way north past Central America, southern Mexico, and the Baja peninsula. The early part of the voyage was "truly a pleasure trip, mountain ranges in sight nearly all the way," she wrote Mattie later, but the farther north they went, the less the Pacific lived up to its name.

By the time the voyage was nearly over, Sara was grateful to pass through San Francisco's Golden Gate, even if in a thick fog and after dark.

Given her fragile lungs, the trip had been hard on Sara. She spent several weeks recovering in San Francisco, later proclaiming proudly to Mattie that she'd gained five pounds. But in typical determined Sara style, she wasn't about to languish in bedrest. Despite a hollow cough every morning, by afternoon each day she felt well enough to head out to explore her new world.

An advantage of being nearly middle-aged was that she'd had years in New York to make connections, and she traveled now with a carefully protected sheaf of letters of introduction. One of her first visits was to Anna Elizabeth (Eliza) Rosecrans, the wife of General William Starke Rosecrans. He was a Civil War Union general who'd led the troops at the September 1863 Battle of Chickamauga (where he'd suffered an ignominious defeat). At first Sara had been reluctant to inconvenience Eliza Rosecrans, but the woman turned out to be so welcoming and warm that Sara wrote later, "I was sorry the stage coach shorted my visit to only about an hour!"

In addition to being tenacious and hyperorganized, Sara was apparently blessed with superlative social skills. She easily made new friends, many of them fellow Unitarians, wherever she went, and she spent much time and effort writing letters and sending small gifts to remain connected to the old ones.

During her several months in the Bay Area, she explored widely, writing Mattie that she had traveled three hundred miles around San Francisco. One day she joined several new friends on the *Vanderbilt* steamer to what's now called Mare Island—not an island at all but a peninsula joined to the mainland by a series of small sloughs. It was originally named Isla de la Yegua, or Mare Island, in 1835 by General Mariano Guadalupe Vallejo in gratitude that his prized white mare swam ashore after the ferry she was on capsized in a storm.

From Mare Island, Sara and her friends traveled to Napa Valley and White Sulphur Springs, the most fashionable destination for the San Francisco elite to escape the coastal damp and cold. Sara reveled in the February warmth—and in the sunburn that lightly blistered her nose, commenting, "Everybody said the burn improved my looks or rather added to it—it did me good."

Despite all the exhilaration and adrenaline of investigating her new environs, Sara was terribly homesick. Observing how ranchers used the western land only intensified how much she missed her father—a farmer—in particular. Micajah Sawyer Plummer served in Captain John Smith's company in the War of 1812, and many years later, in 1831, married Elizabeth ("Betsey") Parsons Haskell. The newlyweds moved first to Portland, Maine, and then later to New Gloucester, Maine. There Micajah took up the grocery business, along with farming on the side. Betsey soon gave birth to Sara's two older brothers, Charles Giddings Plummer, born October 4, 1833, and Osgood Plummer, on April 16, 1835.

The very next year Sara Allen Plummer, named for her paternal grand-mother, was born on Saturday, September 3, 1836—only a month before Charles Darwin would return to England on the HMS *Beagle* with all the biological evidence he would need for his theory of evolution.

Two years later, Betsey gave birth again, this time to Sara's only sister, Martha ("Mattie") Allen Plummer, on August 17, 1838. And, seven years later, making Sara a middle child, Seth Haskell Plummer was born January 8, 1845.

By this time Micajah had sold his grocery store, and he focused on farming full-time for most of Sara's childhood. His faith in the Universalist Church ran deep—so much so that he donated land for a meetinghouse—and both daughters attended Sunday services for their entire lives. He also believed in the power of education, even for girls, and made sure that both Sara and Mattie went to a good private "female" college, the Ladies Collegiate Institute in Worcester, Massachusetts, where Latin and algebra were major subjects.

Betsey's health declined to the point where she was an invalid for most of Sara's life, which might explain why Sara was much closer to Micajah than to her mother. She would write him nearly every week for thirty years.

Now in sunny California, Sara was particularly intrigued by the valley's economics, writing her father that land cost one hundred dollars per acre and grapes were fetching twenty dollars per ton. She saw a variety of investment opportunities—depending on one's gender: "If I were a man, I see many ways that I should delight in making wealth and prosperity crown my efforts." More specifically, she mused, "I should turn my attention to raising stock."

During the same Mare Island excursion, Sara and her group of friends stopped in to visit Dr. Arthur Wesley Saxe, thanks to another letter of introduction given to Sara by her influential cousin Llewellyn Haskell.

Born in New York, Dr. Saxe was one of the many who had traveled overland to California in 1850 in search of gold. That effort didn't pan out for him, yet in 1852 his wife, Mary, and their two small children made the grueling trip to join him—via Panama. Sara's trip was nearly two decades later, but Mary and Sara still chatted animatedly over cups of tea while comparing their journeys across the Caribbean and the Isthmus.

Sara shared much in common with Dr. Saxe as well and described him as "one of the first physicians on the coast, a genius." It's not surprising that she, as an art teacher and future botanist and scientific illustrator, was delighted to meet him: "He sketches the surrounding hills about this Santa Clara Valley (called the Garden of California) and works on an acre of land surrounding his house, cultivates all sorts of rare tropical plants and knows the names and habits of every one and devotes himself to scientific pursuits generally."

A month after her visit, Dr. Saxe donated one of his illustrations, "a beautiful colored drawing of *Rhododendron Californicum*," to the California Academy of Sciences. Eleven years later, Sara would be the first woman allowed to speak to that august all-male group.

It was such a pleasant visit with the Saxes that the group "went back on our way rejoicing" before stopping to explore San Jose. At the time, the town had about twelve thousand inhabitants, and Sara described it as "a beautiful city in nature's wealth of scenery."

In another foreshadowing of her future, they also paused in Oakland, "the Cambridge of San Francisco," where Sara noted the "fine residences and the scientific and literary institutions." Little did she know this community would become home to her and her future husband for many years.

Sara's interests were wide-ranging, and she'd been a music lover and frequent concertgoer in New York. In 1868 she'd made the acquaintance of Ole Bornemann Bull (pronounced *O-lah Bool*), a prominent Norwegian violinist and composer often compared to Niccolò Paganini. Bull frequently toured with Clara and Robert Schumann, Franz Liszt, and others. Coincidentally, he too had recently arrived in San Francisco.

Sara wrote Mattie a week later: "I called on Ole Bull at the Lick House and had a delightful time. He seemed very glad to see me and said, 'O, I remember you.' He showed us his violin, 400 years old, the 'Di Salo,' and I parted, giving him a bouquet of violets. He is full of vivacity and good cheer."

Given all the cultural opportunities, Sara was tempted right then and there to establish her new home in the Bay Area. But Northern California's temperamentally dank winter was too severe for her still-fragile health. She managed to enjoy only several healthy weeks in San Francisco before bronchitis struck once again.

Some years earlier, like Sara, one of her New York acquaintances had also suffered from pneumonia. Thanks to reading Richard Henry Dana Jr.'s *Two Years before the Mast*, he'd chosen to recuperate in Santa Barbara in 1860. On that recommendation, as soon as she was well enough, Sara too headed south by boat for Santa Barbara.

It was a rough trip, and "the little coast steamer tossed like an eggshell, just as one might off Norman's Woe—Father knows what that means—and I for the first time was sea sick." Norman's Woe is a rock reef off the Massachusetts coast near Gloucester, close to where Sara grew up. It's most famous for inspiring Henry Wadsworth Longfellow's poem "The Wreck of the Hesperus," which was first published in 1840, when Sara was about four years old.

Once again, Sara's charm won her the kindness of strangers. The storm had delayed the boat's arrival by six hours, and the passengers didn't land until midnight—when they discovered that the winds had ripped out the old shaky wharf. Fortunately, Sara was a small woman, and the express agent, a Mr. Rosenberg, scooped her up and "equipped with long rubber boots, stalked bravely through surf and sand to dry land, bidding me the while not to be afraid. 'I could carry just such another,' he declared. 'Why, you are as light as a feather.'" Once ashore, he then made certain that Sara was safely settled into a carriage.

After arriving at the American House, a brand-new two-story hotel centrally located at the intersection of State and Cota Streets, she slid out of her own slicker and boots, and collapsed onto the comfortable bed—without bothering to undress any further.

Several days later, on February 13, 1870, she wrote Mattie that the trip had been the worst she'd ever experienced and that her weak lungs had never been so severely taxed. But within a day she was "in good spirits and am up, have a good appetite, and the weather is fine after two days rain."

Fig. 2. Page of Sara's February 13, 1870, letter to Mattie. Photo by author. Original at the UC and Jepson Herbaria Archives, University of California, Berkeley.

Sara wasn't yet convinced that Santa Barbara really was the right place for her. It was still too early to form an opinion of the town that would be her home for the next ten years. She wrote Mattie: "I cannot tell you anything of Santa Barbara, only that it contains about 1,000 people and that it is situated in a valley protected on the north west by high mountains and extending close to the shore. The people seem supremely lazy living in one story a-do-be houses, i.e. The a-do-be is mud dried. Some walls are several feet thick, I am told."

Although her letters never mentioned being afraid, she must have felt some trepidation. Perhaps she even paused before writing, "Now I am at the jumping-off place."

2

"Perhaps You've Heard Our Sadie Was Killed"

Santa Barbara, 1870

↠ AFTER WRITING MATTIE ABOUT her bumpy arrival in Santa Barbara, Sara began settling into her new life. At first her only focus was recovering from the rough trip south. On February 15, 1870, two months after leaving New York, she wrote: "Mattie, I feel ever so much better today, and when I feel worse I will promise to keep my word and tell you all about it. I try very hard to do everything for my restoration. You cannot realize how anxious I am to get thoroughly well. When I allow it all to come home to my thoughts, it overpowers me and it seems very hard to endure. So I drive it all out of mind and do my best."

One remarkable aspect of Sara's health is that she survived so long despite the "medications" she took. During this time she used croton oil, a "preparation of iron" prescribed for her by a San Francisco doctor. It was a particularly foul-smelling and powerful substance often used as a purgative that Sara smeared on her chest to reduce the soreness in her lungs.

She had always been interested in science and medicine—as evidenced by her double certifications in physics and chemistry. After her teacher training at the Ladies Collegiate Institute, Sara moved to Brooklyn where she earned additional certification at the Greenleaf Female Institute. From 1862 to 1865, she'd studied physics and chemistry at the prestigious Cooper Union Institute for the Advancement of Science. One report lists her as receiving a First Grade Award for Analytical and Organic Chemistry—of the eight students, she was the only woman.

That scientific curiosity might well be what led her to teach "calisthenics," or what we now know as physical education. She was also fascinated by medicine, so much so that during the 1860s she spent three years at Manhattan's Bellevue Hospital as a volunteer nurse caring for wounded Civil War soldiers, reading to them and helping them write letters in between changing bandages and bedpans. That hospital experience would prove valuable years later when she'd marry a battle-scarred veteran. Clara Barton, founder of the American Red Cross, was at Bellevue at the same time, and the two women may have met then. Their paths would cross again, resulting in a long-lasting friendship.

But for now in California, some of the company Sara kept could hardly have been healthy. She told Mattie about going for a drive with a distant cousin who was visiting Santa Barbara because his tuberculosis prevented him from working. She concluded she'd rather have her condition than his and that "I think this climate will be good for me."

A month later she reported being nearly as healthy as she had been in San Francisco and said that people were commenting on how much better she looked. But she still chafed at having to be so careful: "I break so quickly that I cannot afford to have drawbacks. Sometimes I fret in the harness it seems such a trouble to try to live." Decades later, she reminisced: "Thus the renewal of life began with this unique experience in the little Mexican-Spanish pueblo of Santa Barbara with its 1300 inhabitants."

The first residents of her new hometown settled what would become the pueblo long before the Mexicans or the Spanish came: The Chumash Indians occupied the area for about thirteen thousand years. Led by the Portuguese explorer Juan Rodriguez Cabrillo, the Spanish didn't arrive until 1542 and, in 1602, gave the area its name in honor of Saint Barbara. El Presidio, the last military outpost constructed by Spain in the New World, was completed in 1792. By the early 1800s the Spanish province of Las Californias was huge: It included what is now Baja California, the state of California, and parts of Arizona, New Mexico, Colorado, Utah, Nevada, and even Wyoming.

After thirteen years of regional struggles against Spain that comprised the Mexican War of Independence, Mexico won the conflict in 1822, and California became a Mexican territory. The Mexican government opened the coastal areas to trade with the United States and elsewhere—

particularly the commodities of tallow and hides, as described vividly, and read avidly by easterners including Sara, in *Two Years before the Mast*.

The year 1846 then marked the beginning of the Mexican-American War, and the explorer John C. Frémont led a battalion of soldiers that took over Santa Barbara. Two years later, in 1848, the Treaty of Guadalupe Hidalgo ended the war, and Santa Barbara became part of the United States. The Gold Rush had already begun, and the town's population soon doubled. Its first newspaper, the *Santa Barbara Gazette*, was established in 1855.

By the time Sara arrived in 1870, the population actually numbered around 2,970, and English had replaced Spanish as the official language. In September of that year the very first telegraph message arrived—from San Francisco. The streets were still dirt, and horses and mules grazed along the grassy edges.

Not just livestock but all agricultural practices caught the attention of Sara, a future homesteader in the making. Even now, just a month after arriving in Santa Barbara, she wrote: "The only feasible thing for anyone to do here who has means is to raise stock, it seems to me so far, or if they could get well located, raise wheat. Anyone with perseverance could take 3000 head of sheep and, reckoning them at $3 per head for good ones here, could at the end of five years clear $20,000."

By now she'd met "Colonel" William Welles Hollister and was impressed by his feat of driving six thousand sheep all the way from Ohio to California. He'd arrived fifteen years earlier in debt by $30,000, but took advantage of buying and selling cheap land. Sara estimated his current worth at probably $300,000.

Her comment was, "I mean to keep my eyes wide open and see all I can."

Some aspects of the American West remain unchanged to this day: Even in the mid-nineteenth century, drought and wildfire were hot topics. In his book *Roughing It*, Mark Twain wrote about inadvertently setting a forest fire near Lake Tahoe "that went surging up adjacent ridges— surmounted them and disappeared in the canyons beyond." In 1870 a conflagration called the Great Fire burned thousands of acres north of San Francisco. A dry spell that lasted from 1862 to 1865 came to be known as the Great Drought, and people were still talking about it when Sara arrived in California. She wrote her father, "Water is the great need here. . . . Five years ago there was a severe drought here & thousands of cattle were driven down from the mountains to a large pen (I passed it yesterday on a drive) and slaughtered for just the hide and tallow."

The frugal, loyal New Englander added:

> The Californian stockholders have never recovered from that disaster, but it might have been avoided in a measure if they had not been so improvident. They never save any of the quantities of grass and make hay in a plentiful time, but every now and then a fire sweeps over the land and destroys all that might be saved in times of plenty.
>
> A Yankee would look out for that, it seems to me. Two crops of anything can be raised here every year.

On another drive, Sara visited the famous gigantic grapevine in Montecito, four miles from Santa Barbara, that was planted in the 1780s. By the time she saw it, "La Parra Grande" was forty inches in diameter and produced twenty thousand pounds of grapes each year. By 1874, though, the vine was dying, so the community raised funds to have sections cut out of its trunk and send them to Philadelphia as a representation of California's agriculture in the 1876 Centennial Exposition.

Not surprisingly, given her personality, Sara soon had a lively social life in Santa Barbara. She made friends with the "very lovely" principal of the local girls' school and went to church every Sunday, often visiting afterward with the Congregationalist minister and his wife. When someone back East recommended a "whiskered" young man of their acquaintance, she responded, "I have not seen the individual that Mrs. Willis mentioned by his whiskers, and strange to say it has never occurred to me since she told me."

Or perhaps her disinterest wasn't so strange—her interest appears to have been elsewhere:

> There was a young gentleman in San Francisco, Ensign Fred
> G. Hyde, a cousin of Dr. Etta's, who is to me the cherished
> acquaintance that I have formed in this country. He is full of
> refinement, kindness, intelligence, and good manners, an ideal
> of genuine manliness that is truly refreshing to meet. He was as
> kind to me as Seth [Sara's youngest brother] would be, and I shall
> always cherish his memory with real affection. He sends me papers
> and every little while sends a word of cheer. On saying goodbye,
> he gave me his photograph which will let you take a peek at—and
> then you must return it in your next. Tell me how you like his face.

Even though that "cherished acquaintanceship" apparently never progressed beyond the occasional word of cheer, Sara's interest in Ensign Hyde was a good sign that she'd moved on from an earlier unsuccessful relationship.

Back in New York in 1868 and 1869, she taught many hours a day at both Grammar School No. 49 and Ward School 14. A promotion to assistant teacher increased her salary to $960 a year. She spent her free time attending the opera and numerous concerts, along with frequent art openings and exhibits—thanks in part to the attentions of Llewellyn Solomon Haskell, who was quite obviously smitten with her.

The relationship with Llew was complicated: Not only was he twenty-one years Sara's senior, but he was also her cousin on her mother's side. And he was divorced.

Like Sara, Cousin Llew was born in New Gloucester, Maine. As a young man, he'd moved to Philadelphia where he prospered as a drugstore magnate. He then married Mary Anna Frost December 4, 1839, when Sara was three years old. Over the next thirteen years, Mary Anna bore him seven children.

In 1852 Llew bought a forty-acre piece of "pastoral" land in Orange, New Jersey, across the Hudson River and thirteen miles from New York

Fig. 3. Llewellyn Solomon Haskell in the 1850s. Photo courtesy of Llewellyn Park Historical Archives.

City. Two years later he built his own rustic retreat there, which he named the Eyrie. Gradually he acquired hundreds more acres, and with the help of his friend and architect Alexander Jackson Davis, he developed America's first gated and planned community. By 1857, Llewellyn Park was advertised as "500 Acres of land divided into Villa Sites of 5 to 10 Acres each with a Park of 50 Acres reserved for the exclusive use of the owners of Sites."

Thomas Edison bought one of those sites and built a twenty-three-room mansion, Glenmont, as a wedding present for his second wife. The Merck and Colgate families had plots there as well. Another house was owned by the Presbyterian minister and abolitionist James Miller McKim and included secret tunnels that were used as part of the Underground Railroad.

Llewellyn Park, with 175 homes on 475 acres, still exists today, and tourists may visit Glenmont.

Records don't indicate when Llew and Mary Anna separated, but census data show that Mary Anna married a second time, to a man named Joel Hall in Fort Wayne, Indiana, in 1865. It is likely that Cousin Llew had been divorced for at least three years by 1868, when he began writing to his "dear Cousin." From then on, he dashed off many more letters to Sara, some in such a hasty scrawl they're impossible to read. From his correspondence, it's clear that Cousin Llew was exuberant, extravagantly wealthy, passionate about the arts, highly intelligent, creative—and besotted with Sara.

In July 1868, during a notorious New York City heat wave, Llew fretted about Sara, and wrote that he hoped she wasn't wilting—which indeed she was. That year the National Weather Service began keeping records in Central Park, and in July the temperatures ranged from 90 to 97 degrees. In that month alone, 833 New Yorkers died from the heat. Sara reported to Mattie that the newspaper accounts were not exaggerated. Sometimes, she said, people died without even having to be exposed to the outdoors: "I have heard of three cases within a week where the persons were well at night and the next day dead." People resorted to all sorts of protection from the sun, including men who walked the city streets with cabbage leaves under their hats—"and palm leaf fans are indispensable to both sexes and all classes." The death toll could have been worse: The city kept public baths and parks open day and night so that people could sleep out of doors. The authorities also distributed buckets of free ice by the hundreds.

But none of those measures helped the nonhuman Manhattanites. Sara wrote Mattie that among the saddest sights were the horses who pulled the city street cars, with sometimes dozens dying daily: "I feel an absolute pity for these poor things even now as I . . . hear the rattling of the hot wheels and the ringing of the car bells."

Sara wanted desperately to escape the city heat, but money was still tight as evidenced by her July 19 request to Mattie: "Can you send me a ticket to Dover?" It wouldn't have been much of an escape: Even Lowell, Massachusetts, recorded a high of 104 degrees in the shade that day.

That torrid summer Sara appears to have resisted her cousin's advances; in the same letter to Mattie, she mentioned casually, "Cousin Haskell called last Saturday with a carriage to take me to Central Park. Of course I didn't see him. And have not."

She was still closely attached to her family, writing often and frequently adding packages of newspapers, books, and plants. One wonders what her mother thought of Cousin Llew, but only one letter from Sara to Betsey has survived. Either Betsey wasn't a letter keeper or she and Sara weren't close, or perhaps Betsey was too occupied with her own medical issues, which the family referred to only occasionally and obliquely. Micajah's rare letters include comments such as "your mother is doing well for her" and "your mother has her ups and downs as ever, at this time she is quite slim."

Sara yearned to hear from her father more often. Perhaps that's one reason an older man appealed to her, and perhaps that's why she eventually succumbed to Cousin Llew's frequent invitations—or perhaps she was simply flattered to have the attention of such a wealthy well-connected man. Or maybe he was just plain fun to be around and had tickets to all the events she wanted to attend—especially the music events they both loved.

Not only were her school days full, but she also spent her evenings and Saturdays teaching classes in painting and how to make wax flowers to private clients. Each week she took part in a ladies' self-improvement group to learn rhetoric and logic—in both ancient and modern texts. In addition, the famous educator and founder of the physical education movement, Dr. Dio Lewis, had hired her to teach his "New Gymnastics" course to classes of both men and women. She also taught an extra class of boys on Saturdays. Sundays were full, sometimes the busiest day of the week, devoted as they were to Unitarian church activities. In between, Sara told Mattie, her "hours were devoted to *sleep*."

She did however manage to see visitors, especially one who made the trip across the river from Orange. In November 1868 she told Mattie, "Mr. Haskell called for me on Saturday to go to one of the Philharmonic concerts, the opening one for the season at the Academy of Music (14 St.). On Sunday I went with him to Steinway Hall (14 St.) to one of [violinist] Theodore Thomas' concerts. It was very fine. On Tuesday this week an invitation to hear [abolitionist] Wendell Phillips, Wednesday to a private theatrical at Mr. Frothingham's vestry rooms (Unitarian), Thursday evening to the opera."

It was most likely during this period that Sara first met the violinist Ole Bull with whom she would reconnect in San Francisco—and from then on, she always referred to him as "Dear Ole Bull."

On January 16, 1869, she attended a lecture with "Mr. Haskell," so whatever reservations she'd had the previous year about seeing him had apparently disappeared. In fact, the following month Cousin Llew dashed off one of his quick notes to Sara saying that even though he'd just seen her earlier that day, he'd meant to invite her to a reception at the Academy of Design:

I entirely forgot this morning! Of course you will go!

Your "Waiting" Cousin, Llewellyn

Sara may have indeed gone to that particular reception, but New York's brutal winter weather—or maybe her overly packed schedule—would soon be her undoing. Mattie, at home in Dover dealing with not one but two ailing parents, must have been disturbed by the letter from Sara's friend, Caroline Whiting, dated March 19, 1869:

Sara took a very heavy cold 6 weeks ago, her lungs felt very sore; she went to see the doctor who ordered mustard. She put on a mustard plaster, soaked her feet, & put on mustard draughts. The next day I told her she ought not to go out, but she said that she thought a ride would do her good, the day was so pleasant. But instead of being satisfied with riding up to 37th St., she went down in the cold yard & used her lungs with her classes; the consequence was that she was laid up. . . .

There is nothing in Sara's case that is worse than in many whom I have known except that it is almost an impossibility to keep her in the house; no matter how urgent the need. She would do the most rash & imprudent things, thinking all the while that she was taking great care of herself, & doubtless using much self-denial that she really deserved credit.

The heavy cold soon worsened into severe pneumonia. Sara's friends, fearing for her life, resolved that she should spend the rest of winter in a warmer climate where she could be safely outside to recuperate. They quickly mobilized, and Caroline Whiting, who also happened to be Sara's boss and principal at their school, procured a leave of absence for her. Sara's friends then shopped for her, packed her trunks, and made all the arrangements for her to join other New York friends who were wintering in St. Augustine, Florida.

Thanks yet again to Cousin Llew, who knew the captain of the ship that carried Sara south, she was well cared for on the journey. During her absence Llew was certainly her most ardent correspondent. By the end of March when she'd been gone ten days, he sounded frantic: "My dear Cousin, I am becoming very anxious about you. I do not hear from you, not even a line." Fortunately the mail arrived as he was writing, and he was able to add: "P.S. Your letter has come at last! Thanks for it, flowers and all, but you must not write such long letters till you are better."

The very next day, he wrote: "O my dear suffering Cousin, how much good your long and interesting letter did me. That ugly cough pains me. I hoped the southern climate would take off the pressure at once, but you need rest."

Somehow, perhaps over the Ward 6 school's spring break, Caroline Whiting was able to take the time to travel to St. Augustine to help care for her ailing friend. Cousin Llew wrote, "You are now as well off as if I were with you, yes, better off. . . . You know somebody that would like to be in your place, don't you?" He also mentioned Ole Bull had invited him to his concert in Elizabeth, a neighboring New Jersey town, but he preferred to stay home, devoting his evening to writing her.

Sara's responses have not survived, but he did refer to a rosebud she sent, so apparently she wasn't entirely discouraging. His letter of five days later further clarified his feelings: "I wish I were with you to take you into my arms and import life-giving strength to you, but that is impossible for more reasons than one. . . . The rosebud! How precious! You ask if your first letter did not arrive. It penetrated to the depths of my being to my inmost soul & produced a calm that would have surprised you. The morbid unrest disappeared as if by magic. My soul is still too full for utterance. I dare not trust myself."

Early that morning Cousin Llew picked some wildflowers to send Sara. Remarkably, they survived the intervening decades and are still tucked among the pages of his letters in the University of California and Jepson Herbaria Archives.

In May he wrote,

Why don't you send some of those [illegible word] kisses that I know you must have hidden away secretly. We will try to wait patiently till your return to take them fresh from you own dear lips . . .

Your Disconsolate but hopeful cousin, Llewellyn

Perhaps Sara rebuffed Llew? In a June 9 letter when he expected her back in New York from the South any day, he wrote, "Hope someday you will find one as devoted to you as I am to [illegible word] on whom you can lean for support at least in the sympathetic direction."

Once she was back in New York, Llew invited her to his house in Llewellyn Park for the weekend with the assurance he'd deliver her to the train early Monday morning in time for school: "I promise to take good care of you and not let you be out in the evening air! . . . I remain that same old cousin of yours."

The letters from Cousin Llewellyn certainly hint strongly at a possibly unrequited love on his part. The relationship may have been another reason Sara decided to escape the Northeast and start her life anew. No other letters between Sara and Llew from that summer have survived.

In fall 1869 Sara fell terribly ill yet again with yet another lung ailment. This time she nearly died. She was now out of options. The decision was clear. She would move across the continent to the warmer, drier state of California even though she knew no one there.

And she would make the move alone.

As she reminisced years later, Sara began her new life in Santa Barbara "with its 1300 inhabitants, dwelling in their gray adobe and red-tiled roof houses, the picture suggesting a symphony of gray and burnt sienna, as they nestled so closely under the shelter of the green foothills, protected by the rugged Santa Ynez mountains stretching in a semi-circle to the north and east."

Early on she began the habit of daily restorative walks, "seeking to interest myself in all people and objects about town. New life was in every turn. The glorious Santa Barbara mountains beckoned; the sunny beach whispered life into my listening ear."

Soon she made the acquaintance of Colonel and Mrs. Bradbury True Dinsmore, of Montecito, a community three steep and treacherous miles east of Santa Barbara (currently known for multi-million-dollar homes owned by Oprah Winfrey, Rob Lowe, Ellen DeGeneres, and more). The

Dinsmores too were New Englanders. Colonel Dinsmore had owned a lumber business among the many that went under in the Panic of 1857. He and his brother drove all their cattle from Maine to California and settled first in Northern California. The climate was too dank for their young son, so he moved the family to Montecito where he bought 130 acres—now the San Ysidro Ranch and Resort.

Colonel Dinsmore is credited with establishing the first orange grove in the state. He also grew almonds, sweet potatoes, bananas, and—Sara's favorite—strawberries. The Dinsmores took Sara under their wing, with Mrs. Frances Dinsmore treating her like a daughter. Sara stayed on their ranch for weeks, building her strength as the Dinsmores plied her with the produce and food raised on their farm.

Sara was even able to get a sample of working with livestock:

> We rise at 6:30. I put on an apron for my dairy work, which is to skim three pans of milk, then go to meet Mr. D. who has a brimming pail of milk in each hand. We have to cross a creek (or as they say here a-ŕo-yo) to reach the cow yard. He sets the strainer pan on a high rock, I sit on another and put lips to pail, and there and then take my fill of warm beverage.
>
> I wipe some of the dishes when I feel like it, set the table, and the balance of the time spend out of doors in the strawberry patch and enjoy all that I can—with a most thorough hankering to be doing something besides this everlasting nursing of myself.

Sara's letters must have sounded all too idyllic to Mattie, whose hands were more than full back in Massachusetts. She worked as superintendent of the Dover schools for nine years and was described as "always an earnest worker in the Unitarian Church." In addition, she was also dealing with her own two young girls and Betsey, whose health continued to decline.

Three months after hearing about Sara's bucolic and serene life, she must have been horrified by Frances Dinsmore's letter that began "Perhaps, before this reaches you, you may have heard that our Sadie was killed—." Several newspapers had published Sara's obituary, reporting that she'd died traveling to the Montecito Hot Springs "while she was yet young and promising."

Fortunately, the media had the facts wrong. Mrs. Dinsmore's letter continued:

> . . . and so knowing how anxious you would feel, I write hurriedly to tell you that she is under our roof and we trust, past all danger.
>
> She is dictating these lines whilst I am writing in her dark room. We were thrown from a carriage, she was very seriously hurt in the head. For three days we have watched over her with very anxious care.

Mrs. Dinsmore assured Mattie that Sara was in good hands ("Trust us, we will take good care of her") and signed the letter with "Give love to everyone from Sadie A. Plummer, alive and likely to get well."

The injury must have been indeed serious because Sara bled copiously from both ears and was incoherent for three days. Even when her "reason" returned, her hearing did not, but the doctor reassured her it would once the inflammation receded. In the meantime, her ears were syringed daily with "sweet oil and brandy."

She did enclose a penciled note to the family that day saying the Dinsmores had cared for her as if she was one of their own and "you may tell Mother that I shall not suffer if I am so far away from home—and if anything should happen that I should not get well, it would be only because it would be an impossibility to save me."

She closed by writing, "I find such good people in the world."

That sentiment was echoed by Cousin Llewellyn, who also wrote Mattie about the marvel of Sara escaping with her life. "Poor Sara! She was doing so well & improving so steadily till this blow came." He concluded, "I shall always have a high estimate of California hospitality & we can be grateful to the many kind friends who took so much interest in her welfare, more particularly Col Dinsmore & his good lady, who have been so untiring in their care of her."

The Dinsmores insisted Sara remain at their house until she was fully recovered. They were getting up in years, and Mrs. Dinsmore told Sara, "It seems so cheering to have a bright young face around the house and to hear your voice. It makes us feel almost young again."

The Colonel agreed, telling her, "If you were our child, we should not do differently by you."

Later Sara wrote home about her injuries and how terrifying to think of her remnants arriving home in a coffin—or even worse, no remnants or coffin at all. When she'd asked Colonel Dinsmore how they'd have sent her remains back to the East Coast, he replied bluntly, "We should have buried you here of course."

It took several months for Sara to recover, and, because she remained partially deaf for many weeks, she had ample quiet time to think about her new life. As she commented to Mrs. Dinsmore after reading her own obituary, what a rare opportunity to discover how one was perceived after death. Perhaps all those words about her youth and promise made her all the more determined to shape her own legacy.

Fortunately, she did indeed heal completely.

In December Sara wrote Mattie, thanking her for the books—and hair rolls (curlers)—she'd sent and reporting she felt as good as new. She also looked back on the previous year, reflecting on all that had happened since she bade the family goodbye as the steamer left New York Harbor. She still yearned to be back home with the family, but she was grateful for what she'd found in California.

Yet she was still restless in Santa Barbara, complaining in particular, "There is not a library or bookstore in the place," and commenting on the great contrast "from the wide-awake life of cosmopolitan New York, with its social and friendly ties to the indifferent sleepy Santa Barbara, for a stranger in a strange land."

Santa Barbara was indeed a far cry from the offerings of Manhattan.

Many years later she reminisced, "My mind often turned to New York's literary centers and their rich storehouses of books; such as the fine Astor library, over which Washington Irving presided for several years. I began to ask myself why we in this far-off region should not have some literary center—with a library attached, if ever so small, to share with the incoming tourist and at-leisure invalids."

So, in typical resolute Sara style, she decided to create one:

I shall try to get up this library if it can be done without detriment to myself in any way—and it may be that it will pay. At any rate it will be an experiment worth trying and will sure to pass the time pleasantly that might otherwise make me feel very uneasy.

You know it is like death to me to be idle.

3

"It Is Like Death to Me to Be Idle"

Santa Barbara, 1871–76

→ NO ONE COULD EVER accuse Miss Sara A. Plummer of wilting on her resolutions.

For eight dollars a month, Sara arranged to rent the rear half of Israel Miller's jewelry store on the south side of State Street between Cota and Ortega Streets. On March 25, 1871, one year after arriving in Santa Barbara, she wrote Mattie that "The Library" had taken all her time and strength—but it was established and already had 350 books. Savvy if inexperienced entrepreneur that she was, she also stocked stationery, artists' materials, and international postal stamps. She added, "This is an experiment that time will tell if it is to succeed. . . . Six months from now I can tell better what is to be. There is not a large field for business here."

A month later she wrote her brother-in-law George, "I can report myself well and established at the head of a little *pretentious* Library." Modestly, she added that she'd send some newspapers to her father that would "express more than my own sense of propriety and knowledge of the facts would warrant."

The *Santa Barbara Weekly Press* had announced:

Miss Plummer, the efficient manager and librarian, with friendly assistance, is putting the Library in order for Immediate use. . . . It is a reasonable source of pride that our town has such a fine library opened in it and Miss P. has complimented the public in general by giving the people credit for literary taste sufficient to sustain such an enterprise. It remains, now, to be seen whether she has paid our public too high a compliment.

We certainly hope not, both for the credit of the place and for the benefit of the fair young lady who has undertaken to establish a good library for the use and benefit of the public.

Once again, Sara's ability to connect with people made the library possible. Dr. Henry Bellows, noted Unitarian minister in New York and one of Sara's friends and mentors, donated and shipped 200 books, with a note saying "a happy thought to take up the work of establishing a public library." She had also lined up subscribers and somehow scrounged another 150 books before the library opened.

Additional volumes came from home: She'd asked Mattie to send the chemistry textbook she'd used at the Cooper Union Institute along with *Plutarch's Lives*, *Under the Willows*, the poems of Jean Ingelow, and Lydia Child's thoughts on old age, *Looking toward Sunset: From Sources Old and New, Original and Selected* (published in 1865).

And, in yet another example of foreshadowing, she asked in particular for Asa Gray's book with the unwieldy title of *Manual of the Botany of the Northern United States, Including the District East of the Mississippi and North of North Carolina and Tennessee, Arranged According to the Natural System*, most likely the fifth edition, published in 1867. Gray was considered the most important American botanist of the nineteenth and early twentieth centuries.

In not too many years, Sara would meet the esteemed Dr. Gray— who would be so impressed with her own botanical skills he would name numerous plants in her honor. But for now her main concern was living up to her reputation as that "efficient manager." She described the library layout to her sister: "The Library is on one side—books all covered in brown paper and numbered, then they are entered on a book alphabetically and numbered to correspond with covers. I have a book for the yearly subscribers and one for transient people. At the rear of this little room 15 ft × 20 is my counter where these daily records are kept and a showcase (cost $20) filled with I.R. stamps, U.S. Postage, pens, pen holders, pencils (Fabers best), visiting cards—white, gray & rose-tinted—purses, port-monaises [a type of wallet], pen-wipers, ink, etc. etc."

At five dollars—more than one hundred dollars in 2020 currency—a yearly library card wasn't cheap: Those who couldn't afford a subscription, probably the "transient people" Sara referred to, had the option of paying ten cents per book for two weeks.

Although she told Mattie that her patrons included "all the best people," she also commented, "Society is a perfect jumble here, but as good in Santa Barbara as probably can be found on the Pacific Slope. What would you think to have a saloon keeper, his bartender and gamblers come in to get books of you, supposing you were the librarian of the Santa Barbara library? They always treat me politely as I try to them, and I think they are generous-hearted people as can be found."

She soon expanded the library holdings to include more art supplies, school books, music materials, and added seasonal offerings such as toys at Christmastime. She even included Valentine's Day cards described by the newspaper writer thus: "Some are exceedingly pretty and will suit the most exquisite taste, while others are comical and extremely ridiculous."

By the end of the year, the library was holding its own financially, and Sara was justifiably proud that she had succeeded in spite of the local expectations:

> This little business takes up my time and I think, after I get fully established and as the place grows, will amount to a nice little business. I do not mean to be too hopeful, and I also move cautiously. I find it is of no use to build hopes very high on anything. So far I can about make the ends meet. It is something to get started. I am pleased with the Library thus far. Everyone said that it could not succeed—they had tried twice before and had failed. I do not mean that it shall, if my health only keeps even with my nerves.

Although far from idle, Sara still yearned for home. She sent her nieces, Sadie and Mattie, each a little paint palette so that they too could paint flowers, and she wrote the older Mattie, "Sometimes I am depressed and homesick and almost think that life is a sickening puzzle." Plus she worried about Betsey, who was ailing: "I'm very, very troubled about Mother's being so ill. You must write and tell me all the particulars. . . . Is she suffering? I do not know that I shall ever see her again—"

The only letter from Sara to Betsey describes discovering a family member—or at least a "chosen" family member. The previous winter Sara had taken the coastal steamer to San Francisco to pick up supplies for the library. When she noticed the captain's name was Plummer, she sent her calling card to him, and he immediately made time to meet her at her earliest convenience. He claimed her as a relative on the spot, and as soon as they reached San Francisco, he sent for his wife and daughter. The three of them hit it off immediately.

Sara and the Plummers remained friends for many years, and she usually stayed with them anytime she was in San Francisco. None of them were able to figure out if they were actually related, as Sara explained to her mother: "I am not good in discovering all the branches of the family tree, to say nothing of the little limbs. I know you are, and if Capt. Wm. E. Plummer, wife and daughter could meet together with the *respectable* part of the M. S. Plummer family, the exact relationship would be discovered, but I cannot do it."

Shortly after this letter, all references to Betsey stop. It appears she died sometime in the fall of 1871. Seventeen years later Sara would refer to their mother's "months of suffering" as she profusely thanked Mattie for her round-the-clock care.

As skilled as Sara was at maintaining relationships, the one friendship that faltered was with the Dinsmores. Perhaps, once she'd established the library, she needed to be closer to town. Or maybe Sara's frequent illnesses made her a high-maintenance boarder? In any case, finding affordable quiet lodgings appropriate for a proper lady was always a challenge. By the end of 1871 she was boarding with the family of E. B. Boust, the editor of the local newspaper, which Sara referred to as the *Santa Barbara Times*. (Confusingly, Boust had started the *Santa Barbara Post* in 1868, and by 1870 it was a daily paper, renamed the *Santa Barbara Press* and no longer the *Weekly Santa Barbara Press* referred to earlier.)

The Boust family lived about a mile and a half from town and even provided Sara with daily transportation: "Mr. B. has a spar of stone gray horses that he drives down in the morning, taking one of his little boys to school, leaving me at the Library at about 9 o'clock, and another boy goes with him to the office of the Times nearby. He puts the horses in a livery stable, and at night we all drive home."

Even though Sara and the Bousts all liked each other and Sara enjoyed the drives to and from work, she reported to Mattie that the four children were very noisy and that she'd soon need other lodging.

Nighttime streets became safer for the town residents on February 21, 1872, when for the first time, new gas lamps were lit. Although Sara herself resisted evening engagements to avoid getting chilled, she had plenty of social life during the day. Her establishment was included in the town's first guidebook, the *Guide to Santa Barbara Town and Country*, published in 1872. The author, E. N. Wood, wrote that the library "is one of the features of the place. Several hundred carefully selected books are collected and new publications are constantly added. Its [*sic*] a pleasant and popular resort among reading people and is something rarely found in western towns of this size." Wisely, Sara catered to the tourist trade as well, stocking Swiss carvings, Chinese and Japanese curios, Santa Barbara stereopticon views, and hometown newspapers from around the country.

Sara was fortunate to have the reassurance and the personal satisfaction of the library—because in 1872, tragedy struck.

The previous fall she'd been thrilled to learn a visitor from home would soon be arriving. She wrote Mattie: "I have just had a letter from 'Cousin' in which he says that he has been very ill with rheumatism and fever, and that when he gets better shall either go to Europe or California and asks what I think of this climate for rheumatism. I shall write for him to come here by all means. I know of a great many who have been cured of that trouble by coming here."

Cousin Llew Haskell did indeed travel to Santa Barbara to visit Sara, arriving in early May. Finally, two and a half years after her arrival in California, Sara was able to see the first familiar face from her East Coast life.

But within a week Cousin Llew fell desperately ill. Grateful for her Bellevue Hospital nursing experience, Sara took charge of his care.

Despite all her best efforts, and far from his New Jersey home, Llewellyn Solomon Haskell died May 31, 1872. Sara was devastated.

"Alone is the word for me," she wrote Mattie.

Throughout her grief, she continued her habit of daily walks. As she wandered long stretches of beach, she became intrigued by the algae on the rocks. Always wanting to keep her family informed about her new life, she created a small book made up of local seaweeds for her little brother, Seth.

By the summer of 1872 she'd met other Santa Barbara women who were intrigued by the local flora and fauna. Several of them, along with Sara, formed the beginnings of Santa Barbara's first natural history association. The other women hungered for knowledge, so, still a teacher at heart, Sara created an illustrated field guide to the algae—little knowing that this type of activity would become her life's work. It must have been popular, as the edition was "exhausted" within a few years.

Soon she learned how to preserve plants, how to evenly dry her botanical specimens, spreading them between sheets of newspaper and blotters, and then flattening them between two rigid boards tied tightly with rope or straps. Once they dried, a week or two later she carefully mounted each plant, arranging it in such a way that the flowers, fruits, or leaf shape remained identifiable from just one side. She then added a label identifying the plant with its collection location, the date, her name, and sometimes a description of the growing conditions: for example, a cliffside or along a stream's edge. Without the information listing where it came from, a plant specimen is useless. Preserved properly, away from insects and from damaging humidity that can rot the plant fibers, specimens last hundreds of years.

Sara's daily rambles soon took her farther afield, "and the mountain slopes were constantly attracting, filling and feeding the mind, thus greatly assisting in regaining strength and vigor. All through the late spring and early summer, I kept up the pleasant recreation of gathering and preserving hundreds of plants for future analysis and study, when the necessary authorities could be made available."

Gradually, Sara became a competent self-taught amateur botanist and illustrator. She studied her copy of Asa Gray's botany textbook, collected plants she didn't know, and sent them off to botanists. To learn about them and observe them even more carefully, she sketched the flowers, the leaves, the seeds. All her works from the time have been lost—except for one. Her earliest surviving watercolor is signed, dated, and labeled "from Nature, Santa Barbara, Cal." The details and the identification of the plant are difficult to decipher because of the brown spots, known as "foxing," a common occurrence in old artworks on paper. Sometimes the spots are similar to rust and emerge when moisture mixes with tiny fragments of metal, left from when the paper itself was manufactured. Other times the blotches are from fungal spores—of which there are more than twenty species that can be found on paper. The specimen, however, is most likely a member of the rose family and might be the blossom of either an apple or possibly a wild blackberry. Later the newspaper referred to a painting of hers that was of an apple tree, so this may have been a preparatory sketch.

Even though Sara had become a sturdy hiker, her health was still fragile: "I couldn't think of going into the school room again. Sometimes my head feels very badly, a tremendous pressure at the ears as though my head would burst if I take a little cold." In August she mentioned being carried home by the physician "with such a violent headache"—which makes one wonder if she was prone to migraines.

Despite her illnesses and homesickness, Sara was falling under California's spell. The climate was "heaven to me": "The weather is very charming. No one ever thinks of saying 'Isn't it a beautiful day?' He'd be laughed at almost, but if it rains, everyone is talking about it."

She wasn't the only newcomer who reveled in the Santa Barbara area. The completion of the new Stearns Wharf in 1872 brought more visitors, some of whom never left. In 1873 a German emigrant journalist named Charles Nordhoff wrote a best-selling book called *California: For Health, Pleasure, and Residence.* (Nordhoff had quite the literary legacy: His

Fig. 4. Sara's earliest surviving painting, signed as "S.A. Plummer 1875. From Nature in Santa Barbara, Cal." Photo by author. Original at the UC and Jepson Herbaria Archives, University of California, Berkeley.

daughter, Evelyn Hunter Nordhoff, was America's first female bookbinder, and his son, Charles Bernard Nordhoff, coauthored the *Mutiny on the Bounty* trilogy.) By 1880, Santa Barbara would add an additional five hundred residents.

The library business fed Sara's intellect but not her wallet, and housing had become either unavailable or unaffordable. By the end of 1872, she had accepted loans or gifts of money from Mattie, her father, and her oldest brother, Osgood. In January 1873 when Mattie asked how her finances were, Sara's discouraged but honest response was still: "I can only report *un*favorably at present" and that she found herself in "a very anxious position."

To supply the library and store, she had borrowed $500—at an interest rate of 10 percent—from Colonel Hollister, the now wealthy rancher who'd driven six thousand sheep from Ohio. She wasn't worried about that particular loan, as it wasn't due for a year. But she was deeply concerned about her day-to-day expenses, since the influx of new residents had driven boardinghouse prices from eight dollars to twelve dollars per week. Worse yet, the building that housed her library had been sold to a sewing machine agency, and she had only four weeks to find new living quarters.

Sara soon calculated that she'd be better off owning than renting:

> On the opposite side of the street is a little wooden building about 10 × 20 ft in size, well-finished. I have bargained for this for $300, and in this I shall have to put about $200 for repairs to make it available for the Library and little store. So I have decided to extend it in the rear 20 feet, and this will give me two rooms to live in.
>
> I can more than save the money by the end of the year in boarding myself.

She broke down her current total monthly expenses of "$60 per month: $48 for board, $3 for washing, and $8 library rent—times 12 months. At the very least $720 per 1 year."

Then she calculated "the future prospecting" of buying the building:

Cost	$500 1st year
Washing	say $40
Interest	$50
Eating @ $5/wk	$240 1 yr
[total]	$830 pr yr

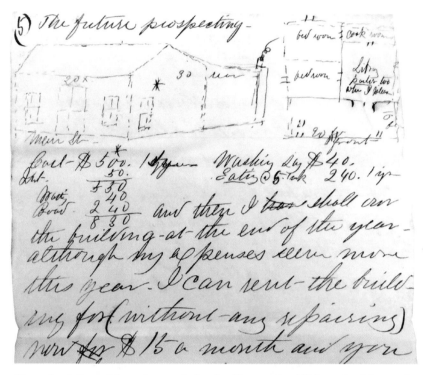

Fig. 5. Sara's financial musings and sketch of her proposed library building and living quarters from a January 27, 1873, letter to her sister. Photo by author. Original at the UC and Jepson Herbaria Archives, University of California, Berkeley.

"And then I shall own the building at the end of the year." She then added a sketch of the future library (see figure 5).

Thanks to California's 1852 community property legislation, even single women were entitled to be "sole traders." Sara's worries about money weren't just for her immediate needs—nor was she planning to remain single the rest of her life. In the same letter, she told Mattie, "How I long to have somebody sometimes to just now lift the burden of life, but if I only can prove that I have the power to hold on, within three years I may become well off enough to feel free of overanxiety. This I look forward to, and hope keeps me up. Two years ago I took out an insurance policy that will pay me back $1500 at the end of 15 years by paying about $40 per annum, so this I shall have for *old age* and my *grand*children."

Fig. 6. Sara Allen Plummer in an undated photograph. Courtesy of Santa Barbara
Public Library.

A year after Sara's musings to Mattie about her purchase plan, a visitor to the new library building described it as "a cozy retreat where you can sit down and revel to your heart's content in the latest novels, poems or magazines. Quiet nooks, cushioned seats and sofas invite you, while the pleasant welcome and greeting you receive from the fair and ladylike proprietress is indeed charming and home-like." The parlor had a small fireplace—and even a skylight. In later letters, she would call her little home "Cactiden," presumably referring to her chosen houseplants.

The library welcomed residents, transients, and tourists alike, and Sara was justifiably proud that English, Spanish, French, and German were all spoken in her store.

Despite the success of the library—or maybe because of it—Sara's health woes continued. In 1875 she fell terribly ill yet again, this time with pleurisy, an inflammation of the tissue that lies between the lungs and the chest wall. The pain was so severe when she breathed that the doctor injected morphine into her side—four times.

But once again, she survived and kept going. As she wrote to Mattie one particularly homesick and depressed Sunday afternoon, "Constant activity only makes life endurable here." Mattie was all too familiar with constant activity: In March 1874 she had given birth to her fifth child, George Osgood Everett. Within two months Edward, her nine-year-old son, died. She would lose George as well, seventeen years later. In all, she would outlive three of her sons.

Aside from homesickness, given the ideal climate, compatible society, and the satisfaction of the library work, most of the time Sara enjoyed her new California life. Yet she still missed the cultural stimulation of Manhattan. So she began setting up lectures, exhibits, small concerts, and social gatherings. Decades later, in March 1910, she reminisced in the *Weekly Press*, "During several successive years, when any public social function was to be projected, the library rooms were where the first gatherings centered."

One of those functions took place in August 1874 when two hundred guests and artists gathered to enjoy a fine display of paintings and a "first step toward the creation of a well-defined art circle in our progressive city,"

Fig. 7. Miss Sara Plummer, undated but most likely in the 1870s. Original at the UC and Jepson Herbaria Archives, University of California, Berkeley.

said one reporter. Sara wore herself out preparing for the exhibition: She even covered all the bookshelves with "unglazed maroon cloth and tin reflectors placed favorably with rows of candles" to illuminate the more than one hundred oils and watercolors lent by local citizens. According to one newspaper account, some of the paintings were Sara's own, one of apple blossoms and another of a large tree in a nearby canyon, both of which have been lost. It was a gala event that called for full formal wear. She wrote, "I was told that several wedding dresses, long packed away, were brought out and gladly donned for the occasion. Everybody was happy and declared the evening a brilliant success."

Two botanically significant events occurred in Santa Barbara in 1876.

According to local lore, that year a visiting sailor gave a seedling of an Australian Moreton Bay fig (*Ficus macrophylla*) to a young girl. She planted it at 201 State Street, not far from Sara's stationery store and library, also on State Street. A year later, that young girl moved away and one of her friends transplanted the seedling closer to the ocean. Today, with a spread of 175 feet, it's said to be the largest Moreton Bay fig in the United States.

The second botanically significant event of 1876?

Santa Barbara's growing reputation as a center for scientific and artistic activity attracted a certain tall and strikingly handsome Civil War veteran, a man who was by now considered to be an important western botanist: John Gill Lemmon.

Fig. 8. John Gill Lemmon, known as "JG," at age thirty-four. Courtesy of Hunt Institute for Botanical Documentation, Carnegie Mellon University, Pittsburgh, Pennsylvania.

4

"A Great Botanist from the Sierras"

Michigan, the Civil War, and Northern California, 1832–76

⟶ IN 1862, A DECADE before Sara established her library, a young man in Lima, Michigan, was choosing to exchange the world of education for that of warfare.

At thirty years old, John Gill Lemmon was, like Sara at the same time, a teacher. Known to his family and friends as JG, he'd graduated from the Normal School in Ypsilanti, Michigan, and then taught in village schools for eight years in his hometown of Lima. He was then promoted to county school supervisor, a job that convinced him to start attending classes at the University of Michigan in Ann Arbor.

JG was a sturdy, athletic young man, musically gifted and fiercely antislavery and pro-temperance. In between his farm chores and school responsibilities, he spent every spare moment exploring the plant life amid the hills and swamps of southern Michigan. Later he claimed to have been born a botanist and wrote that his mother declared that her youngest son had "inherited the re-incarnated spirit of an ancient weed-puller."

And then the Civil War called.

JG was ready to put his life on the abolitionist line, and on August 8, 1862, he enlisted in the Union army. He was assigned to Company E, Fourth Michigan Cavalry, and three weeks later his company mustered and headed to Louisville where they were attached to the First Brigade, Cavalry Division, Army of the Ohio. Known as "Minty's Brigade," the group was led by a colorful Irishman and soldier of fortune, Colonel Robert H. G. Minty.

The company's first engagement was a skirmish against a group of Confederates led by General Braxton Bragg at Stamford, Kentucky,

Fig. 9. JG Lemmon's Civil War photo on which he listed his hospital assignments (as either patient or administrator), the battles he was involved in, and the prisons where he was held. Photo by author. Original at the UC and Jepson Herbaria Archives, University of California, Berkeley.

on October 14. From there they marched 160 miles over eight days to Gallatin, Tennessee, thirty miles northeast of Nashville.

For the next three years, the company marched, camped, fought, and marched again throughout the South, while participating in 102 engagements—36 of which involved JG. Skirmishes, reconnaissance surveys, and expeditions included action at Murfreesboro, Shelbyville, Chattanooga, and Franklin in Tennessee, and Dalton, Georgia, among others. He was hospitalized several times: in Nashville three times and at other hospitals twice. He was so ill in Nashville his older sister, Rebecca Lemmon, who had trained as a nurse, traveled to Tennessee to take care of him. She then remained and worked with him at several different hospitals. Although some of the time he was a patient, JG also worked as a nurse, gleaning medical knowledge that would come in useful later on in his wartime career. Occasionally he served as quartermaster, in addition to preparing bodies for burial.

The spring of 1864 found him back in action in Georgia, and, according to the notes he inscribed on his photograph, he fought May 18, May 24, and June 9 at Kingston, Dallas, and Big Shanty. All three battles were part of the Atlanta Campaign led by William Tecumseh Sherman.

By the time summer's brutal heat and stultifying humidity were crushing North and South alike, JG was fighting at Kennesaw Mountain on June 27 and then at Rosswell on July 4. The soldiers then endured the continuation of the Atlanta Campaign, which dragged on through mid-August.

JG's notes indicate he then fought at both the Battle of Jonesboro on August 19 and Lovejoy's Station on August 20. The goal was for the Union troops to raid the Confederate supplies and destroy their railroad supply lines.

Nine days later, on August 29, 1864, according to both the Fourth Michigan Cavalry records and his own letter, John Gill Lemmon was captured by the Confederate forces near Sand Town (now a suburban neighborhood southwest of Atlanta), most likely as part of the continuing Atlanta Campaign.

He was then transported to the notorious Andersonville stockade.

There was absolutely nothing lucky about being a Civil War prisoner, but, for JG, being captured late in the war rather than earlier probably saved his life. By 1863 one of every five individuals in the town of Richmond, Virginia, was a Union prisoner. Conditions grew so appalling and food supplies so depleted that a new prison camp with the official name of Camp Sumter in Andersonville, Georgia, was established. The 16.5-acre facility wasn't even complete when trainloads of Richmond-based prisoners, already weakened by months of disease and malnutrition, began arriving in February of 1864.

By May the prison had already far exceeded its quota of ten thousand men.

By August 1864 when JG was captured, the population of Andersonville was thirty-three thousand. Most of the men had hookworm, scurvy, or yellow fever and were starving, besides suffering from whatever wounds they'd sustained in battle. Worse yet, they were driven mad by biting lice and open sores crawling with maggots. A prisoner died every eleven minutes.

Weakened by malnutrition and scurvy, JG was soon too feeble to walk. Fortunately, he came from tough stock, particularly on his mother's side.

Amila Hudson Lemmon was a descendant of Henry Hudson, the English explorer for whom the Hudson River and Hudson Bay are named. She was born August 27, 1802, and both she and William Lemmon came from farm families in Geneva, New York. They married when she was nineteen, and she bore six children over the next ten years.

The couple decided to relocate to the Michigan frontier in 1830, where they built a log house they called "Lake Cottage" by Four-Mile Lake in the community of Lima. Amila gave birth to two more children, one of whom was John Gill Lemmon, born January 2, 1832. An eighth child, a girl, was born when JG was two.

After several years, the farm became less productive, and William became a stock drover, traveling frequently for weeks at a time, collecting livestock from Ohio. Malaria, cholera, diphtheria, and typhus were all common, and when JG was three, his eleven-year-old sister died. Then the family lost its primary breadwinner when in 1836 William died of a fever.

William's death left Amila a widow at age thirty-four with seven children under fifteen to feed. Luckily, she was adaptable, hardworking, and quick-witted, so she soon learned to be a midwife and general nurse—occupations that kept her more than busy for the next two decades. She also married her sixty-year-old prosperous neighbor, Justin Baker, perhaps in an effort to ensure her family's financial security.

Gold "excitement" struck the nation in January 1848, and Amila's three eldest sons, Frank, William, and Alexis, soon headed West to join three hundred thousand other gold seekers setting off for California. Cholera claimed Alexis soon after they arrived, but Frank and William settled in California and thrived—despite never striking gold.

Amila's husband, Justin, died in 1852, and eight years later she too left Michigan to join William and Frank in the West. She moved to Marysville, California, near Sacramento, while JG, now in his late twenties, remained in Michigan with his sister Rebecca and brother Charles.

Native Americans were well aware of the potential flooding of the Marysville area, sometimes referring to the Sacramento Valley as an inland sea stretching all the way from the Coast Range to the Sierra Nevada Mountains.

The white settlers didn't pay attention.

The autumn Amila arrived was a particularly wet one, pummeling Los Angeles with rain for twenty-eight days. That fall was followed by an unusually large amount of snow—which, in turn, was followed by an abnormally warm winter spell. The deeper-than-normal snowpack melted quickly and, combined with the heavy rain, led to devastating floods across the Sacramento Valley.

Hundreds of people drowned, along with six hundred thousand sheep and lambs, as well as two hundred thousand cattle. So much rain fell that the oyster beds in San Francisco Bay were covered with freshwater silt.

The deluge left Amila's residence stranded three miles from dry land, but William was able to drive a wagon box across the sheets of moving water to rescue her. Her house was less fortunate: Floodwaters swept it away, along with all her belongings, leaving her with one silver spoon as her sole possession.

After living with her daughter in Carson for a year or two, Amila moved to Sierra Valley in Northern California, to live with her oldest son, Frank. And there she waited for news of JG, her "soldier son."

By early September 1864, Sherman had succeeded in defeating the Confederate forces in Atlanta. In November he headed for Savannah in the infamous March to the Sea during which he and his troops torched every house, barn, and shed they encountered, along with every field of crops. They burned the railroad lines, heating and bending all the rails to render them useless, in some cases even wrapping them around trees where they became known as "Sherman's neckties."

They also freed thousands of slaves. Fearing the release of hordes of Union soldiers along with the freed slaves, the Confederates quickly transferred a group of prisoners away from Andersonville. JG was fortunate enough to be among them, and on September 19, he was moved to a stockade in Charleston. From there he was transported to a hastily constructed prison facility in Florence, South Carolina. Eighty miles east of Columbia, it was conveniently situated at the intersection of three railroad lines.

On November 22, 1864, JG wrote a letter to his family on Florence Prison stationery, and, remarkably, it survived. In it he asked for soap, hams—and a housewife.

By October more than twelve thousand men were "housed" at the Florence stockade. But JG reported to his family that his health was good and he'd regained enough strength to chop wood. Even though he'd attempted to escape—and was chased down by bloodhounds—his jailers awarded him "parole of honor" status and assigned him to be first a nurse, then a steward in Hospital Ward One. They even ordered him to entertain Rebel socials by playing the flute that he'd somehow managed to hang on to throughout his military service—the same flute that would hang on the wall years later at his Oakland herbarium.

He was still strong enough to help prisoners who were worse off by bringing them water in the cup he'd carried for three years of military campaigns. But conditions soon deteriorated, frigid weather closed in, and both food and firewood were scarce—or nowhere to be found. By the depth of winter, few POWs wore more than shreds of clothing. Although Florence may not have been as notorious as Andersonville, just as many men died there, partly because the conditions were equally horrific and because they were already so weak from their time at Camp Sumter.

Probably because he was a more recent arrival to Andersonville—and perhaps because he was his mother's son—JG was stronger than many of his fellow inmates despite having been at war for three years. But once he was moved to Florence, six more months of horrendous conditions took their toll on him, both physically and emotionally. Occasionally prisoners were offered a chance at freedom—but only if they agreed to the Rebel oath of allegiance. JG was tempted a couple of times but couldn't bring himself to abandon his Union principles.

Seeing defeat in their future, in February 1865 the Confederates abandoned Florence, leaving the remaining five thousand Union prisoners to be transported to Wilmington, North Carolina, by train. Of those, only 130 could walk. JG and the rest were then taken by boat to Annapolis where they could at last scrape off the lice, bathe, and step into clean uniforms. Men from the West, which included Michigan, were then sent by train to Camp Chase, near Columbus, Ohio, and awarded a thirty-day furlough. JG went home to Michigan, but within two weeks he was back at Camp Chase, hoping to rejoin his regiment in time for what would surely be the war's last campaign. He apparently weighed ninety-five pounds, and not surprisingly, his request was denied.

In April two momentous events occurred: On April 9 Robert E. Lee surrendered to Ulysses S. Grant at Appomattox, Virginia. On April 15 John Wilkes Booth fired the bullet at Ford's Theatre in Washington DC that killed President Abraham Lincoln.

And then, not three weeks later, Jefferson Davis was captured in Irwinville, Georgia, by none other than JG's own regiment, the Fourth Michigan Cavalry. For the entire remainder of his life, JG wished he could have been part of capturing that "instigator of the wicked insurrection in behalf of slavery."

The train carrying Abraham Lincoln's body passed through Columbus, Ohio, on April 29, on its way to Springfield, Illinois. JG was among a group of released prisoners who marched in a position of honor close to the casket. Two weeks later, Lincoln's coffin would be a part of another procession in New York City, where a Miss Sara Plummer would write her sister about the "terrible national bereavement" and her opportunity to see Lincoln's "calm visage."

By the time the war ended and JG was mustered out on June 9, this once strapping young man was too weak and emaciated to work the family farm at Four-Mile Lake. Based on his "general debility and nervous prostration," he was awarded a 50 percent disability pension of four dollars per month. By this time, before leaving for the West, Amila had sold the farm to the three siblings still based in Michigan: JG, his brother Charles, and sister Rebecca. By spring of 1866 JG decided he too would move to California, and the three siblings agreed to sell the family home.

In late September JG began the journey to California by sea via Panama—the same route Sara would take five years later. Twenty-three days after leaving New York Harbor, on October 14, he disembarked at San Francisco. He then made his way by train and stagecoach to Sierra Valley, twenty miles from the Nevada border, to settle in with family—and to either die or recuperate. He was so worn and fragile that the line between the two options was a fine one. A year later he wrote:

> It was in the mountain home of my elder brother Frank, one mile south of Sierraville, Sierra County, California, that I woke up one early morning of October 1866, an emaciated, feeble survivor of Andersonville prison atrocities, then increased by a liberal diet of one year, to the weight of about 90 pounds.
> As I peered out of the window, and later groped about the premises, the strange flowers, bushes, and even the trees, proclaimed the fact that I was in a practically unknown world.
> Imagine the ecstacy of my re-awakened mentality—a terra incognita, a paradise, an open field of opportunity; all in sight!

JG spent part of his days writing accounts of his Civil War experiences for newspapers. His vivid prose was lively, and his articles were popular. At the same time his interest in botany also reawakened, and, as he slowly healed and gained strength, he began to gather local plants. Sometimes Amila helped, sometimes Frank, and often the owner of the nearby Webber Lake Hotel, Dr. David Gould Webber. Soon his explorations ranged farther afield. Within a couple of years, JG had gained strength, a respectable amount of botanical knowledge—and a collection of mystery plants he couldn't identify.

The nineteenth-century American botanical world was a kind of solar system, a "plant-ary" system with especially prominent botanists including George Engelmann, Sereno Watson, Henry Bolander, Edward Green, John Muir, Charles Parry, Cyrus Pringle, and more—all orbiting around one celestially scientific body in particular: Dr. Asa Gray of Harvard University.

Gray had started his academic career in medicine, earning his medical degree before he turned twenty-one, and then studied under John Torrey, a man indisputably a pioneer of American botany. When Torrey died in 1873, Asa Gray became America's foremost botanist and, after a long career of teaching botany at Harvard, he retired as professor and became the full-time curator of the Harvard Herbarium (which to this day is called the Gray Herbarium). He traveled extensively in Europe, working with many scientists and gaining the financial support of the famous botanist Sir William Jackson Hooker.

His book *Manual of the Botany of the Northern United States, from New England to Wisconsin and South to Ohio and Pennsylvania Inclusive*, known simply as *Gray's Manual*, is still used today.

Perhaps Asa Gray's greatest gift to botany was building and nurturing a network of collectors and taxonomists—two of whom would be Sara Plummer and JG Lemmon. Gray was known as a "severe but encouraging" mentor, which was just what both Sara and JG needed at that point in their separate careers. (JG had a tendency toward sloppy and

Fig. 10. Dr. Asa Gray, photographed here sometime in the 1870s, often called "the father of American botany." Photographer unknown (Mondadori Publishers); public-domain image obtained via Wikimedia Commons.

incomplete labeling of his specimens and particularly needed Dr. Gray's chastisement.)

On his seventy-fifth birthday, in 1885, Gray was presented with a silver vase inscribed "1810, November eighteenth, 1885. Asa Gray, in token of the universal esteem of American Botanists." Including the names of everyone who wanted to participate was a problem, however, so a tray with the names of 185 American botanists was added.

These men of the botanical solar system formed a lively and tight-knit community as they discovered, collected, discussed, described, and named hundreds of American plants, particularly those of the western states. Some were explorers and collectors who scrabbled up and down remote hillsides, climbed prickly fir trees to gather cones from the top branches, and waded through mosquito-ridden ravines. Others worked as dedicated taxonomists, hunched over their desks and microscopes for years, examining tiny details of leaf and flower and deciphering how each plant was related to the others.

And some did both.

Ecology was a concept that was still being defined. These scientists had to identify the biological puzzle pieces before visualizing the picture that revealed why each species lived where it did. They were the giants whose shoulders would support future ecologists.

They were also a remarkably chatty bunch and devoted many hours to writing one another. Letters that flew among them exchanged extensive travel plans, convoluted discussions about whether a plant was new or simply an oddball form, verbal brawls about Charles Darwin's theory of evolution, and the occasional squabble or long-lasting—or even permanent—feud.

But what about women? one might well ask. Weren't women interested in plants?

It was not only acceptable but fashionable for nineteenth-century "ladies" to involve themselves in amateur botany. A 1982 study by Emanuel Rudolph revealed 1,185 women of that time who referred to themselves as active in botany. Some, like Mary Katherine Layne Curran (Kate) Brandegee and Alice Eastwood, went further and became professional botanists. Others remained in partial or total obscurity. Yet the 1880 book *Botany* (vol. 2), by W. H. Brewer, Sereno Watson, and Asa Gray, drew readers' attention to female botanical collectors: "It is a pleasure here to make especial acknowledgement of those who by their contributions have aided essentially in the preparation of this Botany of the State. As the frequent recurrence of their names through the two volumes shows, there are several ladies to whom very much is due. Prominent among these are—"

And Sereno Watson (then Gray's assistant curator and his eventual successor) went on to list eight California women who "have all made collections of value, and with scarcely an exception have contributed new species to the flora of the State," including "Miss Sara A. Plummer." Sara spent much time with several of the women included in the list, and her correspondence mentions most of them.

For many of today's naturalists and gardeners, the names Gray, Engelmann, Torrey, Muir, and Parry are part of plant nomenclature and are far from mysterious: the spiny hopsage whose genus is *Grayia*, Engelmann spruce, Torrey pine, Muir's fleabane, Parry's agave, and many more. Cyrus Pringle, Edward Greene, Sereno Watson, and others may be less well-known now, but they all played large roles in both Sara's and JG's scientific lives.

From 1870 to 1874 in between his botanic ramblings in California, JG tried to return to teaching and became a schoolmaster at the local Sierra Valley school. His war injuries often kept him out of class, but a former student described him as a good teacher, surprisingly gentle for a "grimly bearded veteran," and popular with both children and parents.

On the recommendation of another local Sierra Valley schoolteacher, Professor Eliphalet L. Case, who also accompanied him often on his expeditions, JG sent a package of about fifty plants to Henry Nicholas Bolander, the American West's only prominent botanist, for identification.

Bolander was a German immigrant and, unlike many of his botanical comrades, had a background in education rather than medicine. He continued to straddle both worlds, working at one time as the state botanist for California and later as the state superintendent for public instruction. Yet, despite his two full and separate careers, in 1868 Asa Gray said that no one had done more for California's botany than Henry Bolander.

One of those contributions was providing an entryway for John Gill Lemmon into the world of botany. On February 2, 1872, Bolander wrote

Fig. 11. Sereno Watson, who assisted Asa Gray for years and later succeeded Gray as curator of the Gray Herbarium at Harvard University. Watson named several plants for Sara Plummer Lemmon. Photographer unknown; public-domain image, obtained via Wikimedia Commons, from William H. Brewer, "Biographical Memoir of Sereno Watson, 1820–1892," in *Biographical Memoirs* (Washington DC: National Academies Press, 1979), 50:267–99, https://doi.org/10.17226/573.

Trifolium Lemmoni. sny.
Sierra Valley and other localities
in northern Cal. Dic. 1866.

Coll. by J. G. LEMMON AND WIFE,
OAKLAND, California.

The above *five-leaf Clover* when first recognized
and discovered by J. G. Lemmon as new to the
botanic world, was growing in abundance in
the Sierra Valley region, (1866) and was declared by
Stock-men and farmers to be one of the best
native forage plants of the high Sierras (altitude. 4,500 feet).
(This plant was dedicated to its discoverer by Dr. Serens
Watson, of Harvard University and later published in the Proc. of the
American Academy. XI. 127.) Grows in any soil & worthy of extensive
cultivation. (Plate from Lemmon Herbarium,
5908 Telegraph Av. Oakland, Cal.)

Gray, enclosing JG's plants with these comments: "By mail I send you with this letter a small parcel, containing plants collected by J. G. Lemmon, a teacher in Sierra Valley. His specimens are poor; but still they may interest you. In the future he may do better; he is quite an enthusiast and a good mountaineer; he may be able to find many new plants yet in those mountain recesses."

Bolander went on to explain that JG spent much of his time at Webber Lake with Professor Case and requested of Gray, "In connection with this noble character, I would most humbly ask you to dedicate a species to each of these Gentlemen, if there are any new ones."

Asa Gray did even better, replying directly to "Mr. Lemmon": "Lots of new plants, but don't work too hard. Take good care of your health. The plants will wait for you. I congratulate you upon the change of a rebel prison for a California paradise."

And, to JG's great joy, attached to Dr. Gray's letter was a list of no fewer than sixteen new species, including ones that would carry his name:

Lemmon's clover, *Trifolium lemmonii*
Lemmon's Indian paintbrush, *Castilleja lemmonii*
Lemmon's wild ginger, *Asarum lemmonii*
Lemmon's willow, *Salix lemmonii*
Lemmon's onion, *Allium lemmonii*, and
Lemmon's milkvetch, *Astragalus lemmonii*

Fig. 12. Sara's watercolor of *Trifolium lemmonii*, with her notes that read, "The above wonderful five-leaf clover, when first recognized and discovered by J. G. Lemmon as new to the botanic world, was growing in abundance in the Sierra Valley region (1866) and was declared by stock-men and farmers to be one of the best native forage plants of the high Sierras (altitude 4,500 feet). (This plant was dedicated to its discoverer by Dr. Sereno Watson, of Harvard University and later published in the Proc. of the American Academy, XI 127.) Grows in any soil & worthy of extensive cultivation. (Notes from Lemmon Herbarium, 5945 Telegraph Ave, Oakland, Cal.)" Photo by author. Original at the UC and Jepson Herbaria Archives, University of California, Berkeley.

JG's reaction to this news? "I ran out into the yard, waving the letter over my head, and shouting the new name of each plant as it was reached. Fortunately there was no one in hearing, or I might have been arrested and committed to an asylum."

In between classes—and during the winters of the Sierra Nevada Mountains that he spent at the Webber Lake Hotel, deserted that time of year thanks to eight feet of snow—JG's explorations ranged farther afield. He botanized the Lassen Peak Volcano area in 1873, and in 1875 he and Professor Case spent weeks trekking through the Great Basin area near Pyramid Lake. He kept financially afloat—barely—by supplementing his teacher's wages with playing the fiddle at local dances, selling seeds and herbal medicines, arranging flowers for a Sierraville church, and sending hundreds of specimens to his two primary supporters: Asa Gray and George Engelmann, a German-born and -educated botanist and doctor who'd immigrated to the United States to specialize in western and Mexican plants.

By 1876 JG had become a well-respected collector, regularly providing specimens to experts and herbaria nationwide. His calling card announced him as "Botanist, Lecturer, Microscopist, and Collector in Natural History."

On February 21, 1876, JG attended the regular meeting of the California Academy of Sciences, where he donated specimens of twenty-five species. Later that spring, he joined two prominent botanists, Charles Christopher Parry and Edward Palmer, on an arduous trip botanizing the San Bernardino Mountains together.

Charles Parry thought JG a "good fellow" despite being "excessively nervous & fidgety, does not like to stick to steady work, and likes to make a display of what he does besides being short of funds." Parry himself was no stranger to steady work: After moving from England with his parents in 1832, he studied medicine before shifting his focus to botany, studying under Asa Gray, John Torrey, and George Engelmann. He spent seven years working as both surgeon and botanist for the United States and Mexican Boundary Survey and later was the first to climb Colorado's

Fig. 13. John Gill Lemmon in 1876, around the time that he first visited Santa Barbara. Courtesy of the Archives of the Gray Herbarium, Harvard University.

14,278-foot (4,353-meter) Grays Peak—which he measured, as well as naming it for his friend and colleague Asa Gray.

Much as he liked Charles Parry, JG had his own description of who he himself was at the time: "I neither drink, smoke, chew, swear, play cards, or lose my temper, yet I am not a member of any religious order, I am a zealous Odd Fellow and a Patron of Husbandry. In science I adhere to Darwin and Gray. I am unmarried but not a woman-hater. Mean to marry the best woman I can find—after I can travel no more."

Eng. by E.G. Williams & Bro. N.Y.

Fig. 14. Charles Christopher Parry, sometimes known as "the king of Colorado botany," was the first U.S. Department of Agriculture botanist to serve at the Smithsonian Institution. Image by E. G. Williams and Bro., New York, in the public domain, obtained via Wikimedia Commons.

Little did he know he'd already met that woman and that the two of them would travel many thousands of miles together.

On the way to the San Bernardino Mountains with Parry and Palmer, JG had stopped in Santa Barbara, possibly attracted by the reputation of the local library and literary salon—and perhaps by that of the intriguing Miss Sara Plummer who ran both. On March 23 that same Miss Plummer wrote her sister about meeting the esteemed Professor Lemmon: "There has been a great botanist from the Sierras here for a few days. Dr. Asa Gray of Harvard is a warm friend of his—you cannot imagine such an enthusiast as J.G. Lemmon is. He was at my rooms many times, and we are very good friends, *for botany.*" The emphasis is Sara's.

In keeping with the trend of botanical nicknames, Sara told Mattie that she called JG by his scientific name, "Lemmonia." He in turn had named her "Picea Amabilis" for his favorite tree of the Sierras and the species that inspired him to study botany. The tree's common name is silver fir, and more recent research has removed it from the *Picea* genus and reclassified it as *Abies amabilis*. Sara neglected to mention to her sister that *amabilis* means "loving" in Latin.

In April JG wrote a note to the editors of the *Pacific Rural Press* describing his "grand expedition" to the Santa Barbara area and "its incomparable and most hospitable people." In general, the town's residents and visitors, including Charles Parry and his wife, Emily, who'd spent several months in town escaping the Iowa winter, were thrilled to have a "genuine live botanist" visit them—and the Parrys would eventually become among Sara and JG's closest friends.

JG was also in the Santa Barbara area to collect flower specimens for the U.S. Centennial in Philadelphia, an event that Sara said she'd have to miss because of her library commitments. The social and intellectual life in her adopted town was hopping, particularly with the opening of the new Arlington Hotel where only the "very nicest" people stayed, including George Bird Grinnell, the conservationist, and a Mr. Henry Chapman Ford, a well-known commercial printmaker and painter. Sara was especially pleased that Mr. and Mrs. Ford spent every Saturday evening with her: That year Sara, Mr. Ford, and a few other locals officially founded the Santa Barbara Society of Natural History.

In mid-June both Parry and Palmer urged JG to accompany them again on a collecting trip to the relatively unexplored areas of Colorado. But JG had a different itinerary in mind—and instead returned to Santa Barbara. This time the newspaper reported that he stayed more than a month and exclaimed how JG had accomplished so much gathering, classifying, and preserving of floral samples in such a short time. The reporter continued to gush, "Nothing but the psychic force of an enthusiastic lover of a science would give the poise, nerve, energy, endurance and will to do what he has done."

Before JG headed home to Sierra Valley in July, he was invited to a party at the Arlington Hotel honoring Charles Parry, hosted by Colonel Hollister (by that time Sara had repaid the $500 she'd borrowed from him). According to media reports, "the dinner was spiced with wit and washed down with wine." Two days later the *Daily Press* announced that a farewell reception for both the Parrys and Professor Lemmon would be held "at Miss Plummer's cozy study."

At the reception JG was staggered when his new Santa Barbara friends presented him with a powerful microscope "as a testimonial of their esteem." He proudly described it to botanist George Engelmann as "Smith & Beck's (Eng.) Compound Microscope with 12 degrees of power, for 20 diameters up to 400 . . . accompanied by 'Carpenter's Microscope,' a most exhaustive book, and many accessory implements."

Back in her New York days, Sara had stopped volunteering at Bellevue Hospital in 1863 to attend the Cooper Union. Before she left, her colleagues, knowing about her interest in science, gave her a microscope. In a later letter to Sara, JG refers to the gift as "your" microscope—did she facilitate the present at that Santa Barbara reception? And was it the very same one given by her Bellevue colleagues ten years earlier?

As he made his way back to Northern California, perhaps JG Lemmon gazed at the landscape, thinking about the small, delicate, brown-eyed librarian in Santa Barbara—the woman who would eventually become his botanical comrade and best friend.

5

"My Dear, Soul-Knit Brother"

Santa Barbara and Sierra Valley, 1876–77

⤏ PERHAPS IT'S NOT SURPRISING Miss Plummer and Professor Lemmon were drawn to one another. Both were passionate abolitionists and adamantly against the Confederacy and secession. Both were well-read lifelong students, a pair of terrestrial sponges thirsting for new knowledge and obsessed with learning about the natural world— especially its plants. Each was a natural, kind, and experienced teacher, always delighted to share what they knew with other curious learners. They were both social creatures, and both loved music.

In addition, they both constantly battled severe health woes and yet were equally determined to not let those afflictions get in their way. Knowing that Sara had spent years nursing wounded and traumatized Civil War soldiers in New York must have been reassuring for JG. In turn, she was relieved that he too had worked in wartime hospitals and understood the pressures of constantly worrying and preparing for the next illness. Both were fiscally frantic and constantly worried about money.

But possibly what they shared most was that both "Amabilis" and "Lemmonia" had been alone—and lonely—for a very long time. So, after happily botanizing together in that summer of 1876, it was only a matter of weeks before JG and Sara began corresponding. They reveled equally in written words, and soon they were both picking up a pen to "chat" nearly every day.

The many touching—and lengthy!—letters between Sara and JG reveal a caring friendship that blossomed from sharing the microscopic details of various species to devising additional pet names for one another. Within a month she addressed him as "my dear brother Boaz," and he called her "my botanical sister Ruth"—a reference to the Biblical love story between the widow Ruth and the landowner Boaz in the Old Testament's book of Ruth.

Sara wrote mournfully about the absence of her botanical chums. One enormous challenge for female botanists was not being able to travel unchaperoned. Often, the result was being out in nature with people who were oblivious to the rich environment that surrounded them. She wrote JG about how, during a local walk with acquaintances, she knew all the plants, but instead of joyfully shouting out the identifications, she had to remain silent and suppress her urge to proclaim the names—after all, it wouldn't do to seem "pedantic." She confided in him about her love for botany and her hope that her zeal for it would always remain "fresh & green in heart and mind." She fretted about her botanical inexperience and her timidity in accepting help from the likes of Charles Parry and Asa Gray.

He, in turn, agonized over whether he should remain in botanical exploration or retreat to some other occupation with a more reliable income. Wistfully, he wrote, "Can't we dispense with eating and live on letters? I think I could subsist on just a little soup. (No olives, thank you!)"

And they commiserated about coughing, catarrh, and nervous fevers.

Their correspondence began with an envelope of leaves and seeds. Soon they were shipping each other whole boxes of plant specimens. At one point, Sara wrote that a particularly large box and its long descriptive list "touchingly affected me, when I noted how much time and patience you had spent upon it. You are a 'Magnifica' and no mistake—" referring to the stately silvertip fir, *Abies magnifica*. (The species is now popular for use as a Christmas tree.)

In October, three months after he'd left Santa Barbara, JG was both sick and discouraged. He alarmed her by writing, "My work is done. I am doomed—however it be, so it must go into the judgment."

Perhaps she didn't know him well enough yet to realize he was prone to both depression and dramatic overstatement.

"My dear Brother," she replied immediately. "On the contrary, your work is just before you. It is clear as noon day that your most solemn duty is to do nothing but what tends to help you over this perilous physical trouble."

Sara immediately decided that the Sierras were much too cold for his fragile condition (after all, he did spend every winter buried in eight-foot snowdrifts at Webber Lake) and suggested that he move to Southern California, bringing his mother, Amila, and even his sister Rebecca. "It will never do to give way to a hopeless condition," she went on. "You the brave & valiant soldier boy, one ever on the alert for some chance to help others! You, who have stared death calmly in the face, through battles, pestilence, starvation & imprisonment, must not yield easily now. Dear brother, 'our boy in blue,' California's untiring botanist, my Magnifica, keep up your courage. Ruth and everybody loves you, and you must not leave us for many, many years—"

As one who dealt with her own frequent lung crises, including her recent bout of pleurisy, Sara had strong opinions about how to treat his ailments. Ironically, given her antipathy toward alcohol and her later membership in the Women's Christian Temperance Union, she advised him:

No. 1: Take one pound of white rock candy dissolved by melting it in hot water, or simmer it. Pour it while still hot into one quart of the best whiskey. When it's cold, take a teaspoonful every time you feel inclined to cough.

No. 2: After each meal 1 tablespoonful of the mixture with 1 teaspoonful of cod liver oil poured into it, and swallow without tasting.

Another irony is of course that Sara was in no position to advise anyone on health care. At the time, she too had frequent fevers, and any exertion resulted in an attack of nosebleeds. Nevertheless, she assured him, "I am better and am about every day."

Despite his own financial woes, JG offered to lend or give her money—all of it. Listing his all, which included one hundred "priceless letters" (presumably from Sara herself), he confessed that "as a business man, I am a shameful failure."

Sara thanked him effusively but said she couldn't be selfish enough to accept because, after all, he needed his money as much as she did.

A month later Sara wrote Mattie that she had "a leaky roof, debts, overwork, pleurisy, a meagre larder, homesickness, etc.," but her friends Miss Seely and Miss Whiting had sent a package containing a new dress pattern, along with a piece of black cashmere, another of white linen, and a cardinal-red necktie, along with a pair of scissors to create it. It came in the nick of time, as she had just concluded that she

> needed some fixing up in the way of substantial dresses—and no mistake. My entire & available dresses being an old black silk, a gray cotton & wool bought nearly two years ago, a blue flannel 3 years old but re-dyed, and a cotton & wool 4 years old and re-dyed navy blue and made over this season. "Well," I said. "It cannot be helped. I cannot afford anything new this season, the times have been so dull and hard. I must save all I can, get entirely out of debt" . . . when lo! As if for a reward for my using my common sense, this dress came!

She also wrote that whenever she got discouraged and "overdisturbed" by debts and overwork, "I take up my Botany for an hour, or try my hand with the brush to paint a flower from Nature, or read of the struggles of some wonderful discoverer."

The next year, 1877, was a turning point for the couple, both personally and scientifically. By then JG was thoroughly smitten with Sara. On April 18, 1877, as those eight-foot, "worthless" snowbanks melted into Webber Lake, he wrote the following poem:

"RUTH AMABILIS"

By "B.M." [probably Boaz Magnifica]

The South land holds a prize—*chère amie,*
A lady small of size—*chère amie*

O Sun! Gleam for her.
O Moon! Beam for her—
But with large and flashing eyes—*chère amie*
A city owns her worth—*chère amie*
The scripture salt of earth—*chère amie*
O winds! Blow for her.
O Streams! Flow for her—
Of good and noble birth—*chère amie*
The poor have there a friend—*chère amie*
The sick a nurse t'attend—*chère amie*
O Hearts! Warm to her,
O Souls! Form to her—
And the crushed a priest to mend—*chère amie*
The Classics early sought—*chère amie*
Then Art her fingers caught—*chère amie*
O Sage! Lend to her
O Muse! Bend to her—
Now Nature wins her thought—*chère amie*
O lovely sister, mine!—*chère amie*
My soul is knit to thine—*chère amie*
O Earth! Spare for her
O Heaven! Care for her—
Shall true arms intertwine?—*chère amie*

Sara couldn't accept that JG might actually be serious and assumed her "Funny Man" was just trying to cheer her up. She sent some of his letters to Mattie, adding, "Prof Lemmon thinks me a pattern of everything ideal and altogether lovely—thinks he has discovered wings—but he is a great enthusiast as you can see. Single hearted, the soul of goodness—a poor Union Andersonville soldier, a physical wreck from war's cruelties, but he has a mind & heart that can never die. You see that he is very susceptible to trifles of kindnesses & sometimes he effuses. Then I call him 'The Funny Man.' Will you return them after reading so that I can answer."

She did, however, allow herself to admit that she was pleased by his attention: "Now this is an illustration to you as to how the sun rays into my heart occasionally."

Sometimes she had to chastise JG, gently, for writing letters that were so personal, so intimate that she couldn't share them with anyone: "You roguish brother!"

Sara was curious about every element of the natural world, and in May she thanked JG for sending her a book about beetles. She told him about being given "a curious marine animal. . . . I kept it alive by frequently feeding it with new ocean water, then I made a pencil sketch of it as it lay in the water and wrote out a description of it." Her friend Colonel Ezekiel Jewett, a noted geologist and "conchologist," told her he'd never seen anything like it, and he wrote an introductory letter for her to enclose with the sketch and description to Spencer F. Baird—who in 1850 had become the first curator of the Smithsonian Institute. Sadly, there's no available record of her letter or even what the creature was.

But her discovery of another organism did survive: The previous summer while exploring a streambed in the Glen Loch ravine in the mountains above Santa Barbara, she'd encountered a six-foot-tall bushy shrub with flowers that resembled fluffy white puffballs. She didn't recognize it, collected a sample, and mailed it to JG. He couldn't identify it either, so sent it off to Asa Gray, who responded that it was a new species.

JG then asked Gray to name it for Sara, exclaiming, "Her name will honor the science!"

Sara was enormously pleased. To this day Plummer's baccharis still bears the scientific name of *Baccharis plummerae* in Sara's honor. Its official botanical description includes Asa Gray's comment, "discovered by Miss S. A. Plummer, an ardent botanist, whose name it is a pleasure to commemorate."

That same month Asa Gray named a new genus of borage *Lemmonia californica* for JG (now *Nama californica*, common name California fiddleleaf). In the scientific description, he wrote: "Of late years I have had frequent occasion to associate the name of Mr. J. G. Lemmon with species of his own discovery; and I seize with satisfaction the present opportunity of further commemorating the services of a most ardent and successful explorer of the Sierra Nevada region, by naming in his honor this interesting new genus which he alone has met with. By the specific name, Californica, I indicate the principal field of Mr. Lemmon's arduous explorations."

The 1870s and 1880s were lively decades for botany for both men and women. The growing seasons would find Parry, Pringle, Greene, Engelmann, of course JG and Sara, and all the other collectors out scouring remote canyons, ravines, and cliffsides in search of new species. Competition was intense, and more letters exchanging species debates, as well as personality tidbits and even gossip, flew among them all.

But in 1877, a remarkably collegial international camping trip occurred. In September, after what must have been an astonishing amount of transoceanic planning, the famous botanist and explorer Sir Joseph Dalton Hooker visited the United States. He was the director of London's Royal Botanic Gardens, Kew, and the closest friend of Charles Darwin, with whom he exchanged fourteen hundred letters. The founder of geographical botany, Hooker had a most heartfelt goal of traveling with Asa Gray through the American West while exploring and collecting amid deep and passionate discussion with other botanists. JG told Sara it would be an expedition taken by "the chief botanists of both hemispheres!"

While in Northern California, Gray and Hooker realized they could also spend a few days camping and exploring with John Muir—after all, who knew the Mount Shasta area better than a naturalist, mountaineer, botanist, geologist, and glaciologist, all wrapped up in that one man?

In 1940, Andrew Denny Rodgers III, a botanist who wrote several books about the history of botany, described the botanical get-together: "One September evening, encamped on the flanks of the mountain in a forest of silver firs, a log fire was built and storytelling began. Gray told of his explorations in the Alleghenies; Hooker of his in the Himalayas; and they talked of trees, arguing relationships of various species, and Sir Joseph admitted to Muir that 'in grandeur, variety, and beauty, no forest on the globe rivalled the great coniferous forests of [Muir's] much loved Sierra.'"

What living naturalist with a knack for time travel wouldn't drop everything to be sitting at that campfire!

Sadly, owing to their health issues and Sara's library responsibilities, neither JG nor Sara was available for that particular trip. But JG was able to meet both Asa Gray and Sir Joseph—thanks to Sara.

By now, due in part to her own curiosity, determination, and work ethic and in part to JG's tutelage, Sara was an accepted botanist in her own right. She had gained the confidence to correspond directly with Parry, Engelmann, Greene, Watson, and Gray, and she was also beginning to collect and exchange specimens with botanists in other countries. Even though she wouldn't be accepted as the California Academy of Sciences' second female member until the following year, she frequently traveled to San Francisco to work in the academy's herbarium, which is where she was the day Dr. Hooker arrived.

After meeting the "truly noble" Sir Joseph Dalton Hooker ("who's none the worse for" his recently bestowed title), she wrote JG "he looked at me an instant & with a most cordial shake of the hand said, 'O, I have heard Dr. Gray speak of you. He also spoke of having a letter from you with an invitation for us to visit Santa Barbara. And I regret that I am the one that must needs prevent the party accepting your kindness, but my eldest daughter was just married as I left Kew. I promised to join her & visit them before the honeymoon should be over.'"

They continued chatting, and he suggested they meet again later that day at the Palace Hotel so that she could also talk with Dr. Gray. Thrilled, she of course took him up on the invitation, and, after visiting in the hotel with him again ("He entertained me there for over half an hour!"), "then in came Prof Hayden, Mrs. Gray & lastly charming Dr. Gray. He drew his chair beside mine and told me just why they could not visit Santa Barbara. Said that he feared they must leave you out. Then I said, 'Dr, you must in some way arrange to see our good soldier boy & the most enthusiastic botanic worker on the Pacific coast—& your great admirer.'"

A day later Sara wrote JG with the exciting news that she'd had a "long talk" with Gray and that the two eminent scientists could meet with him for a day of botanizing. That same day Gray wrote to JG, suggesting that he, JG, and Hooker meet at Truckee. From there they would then travel to Lake Tahoe together, all at Gray's expense.

The timing was ideal. Just the day before, JG had written despairingly to Sereno Watson, who was then working at Harvard as Gray's assistant, "My health is very poor. Suffer with a cough & headache. Nearly unfit for anything. If I recover, I will hibernate at Webber Lake again" for the winter.

The news of the esteemed visitors perked him up, and now he desperately wanted Sara by his side to help welcome "the savans," the "eminent personages": "None other could do it better, and none other do I desire."

Sadly, Sara felt she couldn't risk the physical strain, and she remained in Santa Barbara, writing, "I hope your heart has been made glad by sight of them. You must tell me all about it."

The Hooker-Gray expedition did indeed make it to Truckee, where JG had a four-horse coach waiting—and where he was, of course, accompanied by his mother, Amila, who charmed the entire party. The ladies retired to the hotel, and the men spent the day climbing trees to gather cones, whacking off chunks of pine bark, and scrambling around the hillsides collecting as many plants as they could cram into their portable presses in two short days.

For JG, it was a marvelous, worshipful occasion. As he wrote Engelmann, "I had a glorious time of course! Though all the while very stupid with head-ache and lack of sleep, caused by catarrh, happily checked entirely by now. But I was thoroughly afraid of the great men and constantly stammered in discussing the most familiar plants."

All those hours Sara was spending at the California Academy of Sciences herbarium paid off. The minutes of the November 19, 1877, meeting include the proposal for inviting the organization's first two women: Mrs. J. H. Sargent and Miss Sara A. Plummer.

In the meantime, Sara and JG's friendship was continuing to blossom despite the more than five hundred miles between Sierra Valley and Santa Barbara. In December JG finally worked up the courage to propose to Sara. He wrote, in part, "Won't you love me a little & let that little leaven the whole lump?"

Several tortured days of silence followed.

At last, in a lengthy letter written over two long, stormy Santa Barbara nights, Sara refused him. At first she claimed she wasn't sure she loved him—or was even capable of loving him enough. The third morning, after yet another sleepless night ("went to bed last night, dear friend, in a tempest, outside and within, was chilly and sleepless till near morning, so of course today am quite unfitted for what is before me"), her practical side came to the fore:

> The subject is a very knotty one, and there are many practical points that must sometime be faced. . . . It is most conclusive to my common sense that two unhealthy persons should not come together. Let the love be ever so strong, one at least, should be in excellent health. It is terrifying to contemplate! Think of children whom we may see daily, toppling about, born without lungs, liver, or brains! . . .
>
> How sad & disheartened I feel!

This was a heart-wrenching decision for her as she had come to realize she did indeed love him:

> I only know this, that when with you conversing in the sciences together, the blood fairly dances in my veins & I'm all aglow from floor to ceiling—my heart goes out toward you so that I could almost embrace you. My heart seems to expand and I feel so happy—but, there perhaps this is only the natural glow of enthusiasm. I know not. Then again if I could assist you in any way, I'd share my last crust—or dollar. . . .
>
> The embers are low in my little contracted hearthstone, all covered with ashes, smoldering and dying—the lamp's chimney is cracked and the light flickering. I hear the poor ghost of my love moaning without, my sadness is borne to me on the wings of the wind, my soul is bowed in heaviness. . . .

I fear that I shall become the unhappiest of mortals in the remorseful regret.

She signed it, "—a sweet goodnight with peace like a downy pillow for your dear head, and may no more unhappiness continue to grow from the poor loving heart of Ruth."

6

"Into the Matrimonial Vortex!"

Santa Barbara and Oakland, 1877–80

↦ SWEET GOODNIGHTS AND DOWNY pillow notwithstanding, Sara's rejection devastated JG.

Charles and Emily Parry had become good friends with both Sara and JG, and, as he waited in suspense to hear back from Sara, JG wrote them that he feared there would be no wedding. Emily replied, warily, about both the fiscal and physical challenges they'd face if they married: "Few persons have taken a stronger hold on our affections [as you and Amabilis], and the consummation of your hopes and wishes would give us great joy were it for the best—but under the circumstances I fear there might be some clouds to dim the matrimonial horizon. You are both of delicate organization and, without a competence to depend upon, you might see some dark hours though seasoned with genuine affection. There might be trials."

Evidently Sara and JG agreed with her, however reluctantly, and they resigned themselves to remaining long-distance pen pals. Few of their 1878 letters to one another have survived. JG gave up being snowed in at Webber Lake and spent the winter working at the California Academy of Sciences in Alameda, probably to be closer to Sara when she stayed with her "chosen" Plummer cousins. He then lined up work as botanist on a summer-long expedition with Dr. Dio Lewis—most likely with Sara's help, since Dr. Lewis had hired Sara to teach gymnastics twenty years earlier. But Lewis fell ill, and the excursion wilted.

January 21, 1878, was the day the California Academy of Sciences officially accepted its first seven female members, including Sara. JG was also accepted as a resident member at the same meeting.

Whenever she was in the Bay Area, Sara continued working at the Academy, unpaid, sending out plant specimens to collections all over the country. In one package of oaks to George Engelmann, she added a note: "If there is anything in the way of assistance by collecting in this region, it will give me pleasure to forward it to you. . . . I am on the alert at every opportunity offered to collecting and study in our beautiful & exact science of botany."

In June JG returned to Santa Barbara to lecture on the California pitcher plant (*Darlingtonia californica*), and to botanize with, of course, the town librarian. That same month he joined an expedition to Yosemite. Sara planned to accompany him but at the last moment was too sick to go.

The year 1879 started auspiciously for JG. Railroads had expanded their network, allowing collection throughout more of the West, and Asa Gray negotiated with their presidents for free passes for botanical researchers. On January 22, 1879, he wrote JG, giving him permission to "use my name to help get free Railroad Passes for botanical exploration." Expense-paid travel proved to be a huge boon and an arrangement that would benefit both Gray and JG, as well as Sara, for years to come.

And, finally, in spring, JG persuaded Sara to visit him in Sierra Valley, with the added inducement of a collecting trip to Pyramid Lake, Nevada. They would be chaperoned, of course, by Amila Lemmon.

It was a lengthy trip, four weeks, with the three of them jostled together in a two-horse covered wagon. JG gave full credit to Sara when he submitted the eventual list of hundreds of specimens to Harvard: "Plants collected from Sierra Valley to Goose Lake Oregon via Pyramid Lake, Surprise Valley, the Humboldt desert & headwaters of the Pitt River, through Washoe and Roop counties of Nevada & Plumas, Lassen & Modoc counties of California. By J. G. Lemmon & Miss Sara A. Plummer."

Sara loved every moment of it, writing to her friend "Papilio" (possibly Emily Parry) about the rugged, hot, twenty-mile drive over alkali desert before reaching Pyramid Lake, where "our toil & fatigue was for the time forgotten as we reached the summit pass that for the first time allowed the eyes to gaze with delight upon its distant shore—the blue waters as smooth and blue as the sky above reflecting the mountain, rocky islands and the striking rock pyramid rising 600 feet out of the water."

Not surprisingly, Sara was not one to sit inertly as an unoccupied passenger:

> As we went down the grade, in many places the steep and sandy road, narrow, sometimes quite blocked with stones—you should see the developing skill of "yours truly" in handling the ribbons [reins] & managing the brake. The pair of horses drawing our covered camp wagon, heavily loaded bearing also a precious burden—J.G.L.'s mother, over 76 years of age, lively & most entertaining—were not to be trusted for a minute, their feet becoming tender from the long journey through alkali soil & hot sand & stones, caused them to search for easy footing—no matter what the consequences might be at the rear. But we managed to make the descent in safety—at the same time to observe the striking peculiarities & wondrous beauties of this lake.

Pyramid Lake lies forty miles north of Reno and is 120 acres of mildly salt water, about 15 percent that of seawater, and fed by the Truckee River. It's now surrounded by the Pyramid Lake Paiute Reservation. But back in 1879 the reservation was smaller and located about thirty miles from the lake. The name "Paiute" is a historical label that actually designates three groups of Great Basin Indigenous people: the Southern Paiute, based in northwestern Arizona, southern Utah, and southern Nevada; the Northern Paiute, who settled eastern Oregon, southwestern Idaho, western Nevada, and northeastern California; and the Mono Tribe of the Sierra Nevada and central eastern California.

Traditionally the Northern Paiute were nomadic and depended heavily on the lake for fishing during spawning season. Their tribal name is Kuyui Dükadü, for the Lahontan cutthroat trout, or "Cui-ui-Fish-Eaters," honoring the sucker fish, *Chasmistes cujus*, now an endangered species and found only in Pyramid Lake. The land for the reservation had been set aside in 1859, but as the white settlers moved in, conflicts over scarce water resources and land use heated up, leading to the 1860 Pyramid Lake War, the establishment of the Malheur Indian Reservation, and the Bannock War of 1878.

The Plummer-Lemmon expedition met up with Northern Paiute, Sara's first exposure to Indigenous people. Unfortunately she, like many of her contemporaries, regarded anyone who wasn't white as lesser beings. By that time, many Native Americans were understandably wary of whites and stayed on the reservation. Still, Sara, JG, and Amila all felt some trepidation:

> We lay down that night upon our camp beds with some watchful, nervous sensations & every time the willows swayed & sang their willowy strains, I for one thought of the sly tread of these wily creatures. It sounds like the parting of limbs & [illegible word] by a hatchet for the work of a tomahawk. Nothing occurred, however, but next morning before we were up, one "big injun" started into camp. Being politely told to "vamoose," he disappeared.
>
> Later a tattooed squaw with her painted papoose tied upon her back against a board that looked like a small coffin cover, came paddling along with a little boy about 8 years of age strutting in her wake, bow and arrows in hand. We thought the neighborly feelings might have been stimulated somewhat through the olfactory nerves for he, she & it made a halt just in front of our coffee pot & frying pan that were filled and steaming & sizzling over a blazing fire. Now we spoke. She chuckled & sniffed, rapidly unstrapped her papoose—but O, the wave and fatal word "vamoose" suddenly caused her to gather up and retreat with a snarl, grunt & a farewell scowl.

The Pyramid Lake trip gave Sara a tantalizing taste of what life as Mrs. J. G. Lemmon might be like. She wrote, "How I wish that I could impart to you and all friends some of the pleasures of this nomadic camp life."

The expedition also confirmed JG's belief that Sara was the partner he'd been hoping for. He wrote Asa Gray that "Miss Plummer proves an excellent explorer with keen vision & skillful hand. In case some suitable things in our collection prove distinct, I strongly urge that she be honored—both in recognition of service now, and for several years of laborious but unrecognized work in Southern Cal."

The esteemed Dr. Gray apparently agreed with JG. In May he published the official description of the new species Plummer's baccharis (*Baccharis plummerae*) in the *Proceedings of the American Academy of Arts and Sciences*, the one in which he described Sara as "an ardent botanist, whose name it is a pleasure to commemorate."

Life in Santa Barbara apparently had lost some of its appeal for Sara because she visited the Plummer cousins in Alameda—not far from JG's herbarium, which also happened to be in Alameda—more often and for longer durations. One day a small yellow bird dropped dead at her feet on their porch. She didn't recognize it, so she skinned it, prepared it, and sent it off to the Smithsonian, where it too turned out to be a new species. The yellow-rumped warbler is known affectionately among birdwatchers as "butter-butt."

Later that summer Sara made a long-awaited trip back East for several months to visit her family and friends, her first visit home since leaving for California nine years earlier. As it's quite likely she was delivering plant specimens to Asa Gray, she may have traveled on a free pass.

In July Captain Plummer wrote to let her know that JG wasn't doing well in her absence: "Mr Lemmon staid [*sic*] a few days with us. I think him very much more worn than ever I saw him, and certainly more unbalanced. I fear, Sara, that it will not be very long before he will be totally unfit to take care of himself. I have not heard it from him but from other sources that he has procured his pension, at fifteen dollars a month." In today's currency, that would amount to about $360 per month—hardly enough for one and certainly not for two hardworking botanists.

Sara's health and finances weren't any better than JG's, and in September Asa Gray wrote JG, "I fear Miss Plummer is ill. She was here a very few days and then disappeared. Her sister said she had been poorly."

By now, even though they'd never met, JG and Mattie were also botanical pen pals. Mattie, like so many nineteenth-century women, was yet another budding amateur botanist. In October JG sent her a cigar box of seeds, bulbs, and cones with a note referring to Sara as "my little botanical sister Amabilis."

During Sara's absence, JG was still hunting for work and desperate to the point of selling dried-flower arrangements to tourists at Yosemite. He'd applied to join the Fortieth Parallel survey and was crushed when told they didn't need a botanist. George Engelmann had been buying plant specimens from JG, who was so disheartened in November, he wrote Engelmann, "Money has never been so scarce here."

He then announced he was moving his personal herbarium—and Amila—from Alameda to the "wealthy and aristocratic city of Oakland. . . . If I was receiving a salary from Govt as I deserve to, being a victim of Rebel hate, I could get along perhaps, but I expect to go down in neglect and poverty. . . . I have only myself to blame—I might have taken the Rebel oath and gone out of Andersonville."

Another friend of Sara's wrote her, "Prof. L is both lonely and disgusted. What he may do in his desperation I cannot imagine."

After Sara returned from the East, she continued her visits to Alameda, still conveniently close to JG's relocated herbarium in Oakland. The two botanical comrades spent enough time together that in January 1880 Sara wrote Engelmann, "I am glad the *Pinus ponderosa* var Jeffreyi with cones reached you. We took great pains to get it."

That same month she wrote home that she was reading William Prescott's *Conquests of Mexico and Peru*, which she described as "wonderfully interesting . . . more interesting than any novel." In passing she mentioned that Mr. Lemmon had visited and asked her to tell Mattie that the Parry's lemon lily bulb he'd sent would flower in spring "after its bulbous rest." It's a particularly showy and unusually fragrant lily and now considered rare, perhaps and ironically thanks to well-meaning enthusiastic botanists. Earlier that summer JG wrote Sereno Watson that he found a secluded valley inhabited by the lily and had harvested three hundred bulbs. He then asked, "Can you help me sell them?"

In the spring of 1880, illness struck Sara down—again.

While visiting Alameda she'd delivered a botanical talk to the Oakland literary society, despite a terrible cold. After her presentation, the hosts, Mr. and Mrs. Franklin Warner, insisted that she avoid the night air and remain with them overnight instead of starting the journey home. She accepted, and by morning her cold was worse. A doctor examined her at the Warners' house, and announced a dire diagnosis: Sara didn't have a cold; she had spinal meningitis. Heading home to Santa Barbara was out of the question.

Once again ("the way I came here was providential"), Sara had fallen into remarkably kind and supportive hands. Six weeks later, she was still with the Warners, and on April 15 she was finally able to write her family:

> Am propped up in bed but mean to write you a line to tell you that I am slowly on the mend. The Dr. and all who know say it will be months before I become *entirely* well again but that I will, with care. I have a hint that it is a slow process for when I attempt to test the *moral* courage of one set of nerves, for instance the feet, to see how much locomotion they can stand, the heels refuse to touch the floor, or perhaps the back of the head suddenly weakens, or the right arm drops useless, knees give out, eye won't see, ears won't hear, or worse for woman, the *inferior maxilly* [*sic*—part of the lower jaw] suddenly refuses to perform its normal flexible function. I haven't yet been affected in the tongue, and that is the sole exception, but I'm out of the worst of it.

The Warners, both former New Englanders and, according to Sara, "both finely educated and cultivated," spared no pains or expense on her behalf. A week later she'd recovered enough to continue her letter on "a sad & sickening day for Oakland & vicinity" when Berkeley's Giant Powder Company exploded: "There passed by our windows eight horses at one time & six at another bearing the fragmentary bodies of the men who were blown to atoms. 25+ at the Giant Powder works in the adjoining town of Berkeley. I never saw anything sadder."

In the meantime, that spring JG had headed off to botanize southern Arizona. He still had a tendency toward sloppy labeling and had been chastised—yet again—by Gray for not including enough location data in his collections, and this time he was determined to do a better job. Sara knew how much the trip meant to him and was equally determined to hide her latest illness to avoid distracting him.

Mattie, however, shared the news—and severity—of her sister's latest illness with JG. On April 26, JG responded (he'd still never actually met Mattie) from Tucson after botanizing in the Huachuca Mountains: "And all the while I've been gleaning here with eager hand and delighted senses, our dear little sister (whom I call Amabilis) has been suffering sickness near death. She is so careful not to distress her friends that she kept the facts from me all these weeks." He went on to refer to the excursions he'd enjoyed with Sara as "blissful."

Around this time JG learned his friend John Muir had married Louisa Strentzel. One might detect a tinge of envy as he congratulated him for joining forces with the only daughter of a millionaire, writing, "Now our dear little Johnny Muir is in clover. Did ever such good fortune come to a scientist before!"

By fall, California was honoring the thirtieth anniversary of its admittance to the Union, but Sara felt far from celebratory. She was in San Francisco searching for work and discouraged after climbing "many pairs of stairs into offices, editors' dens, etc. to get writing and copying to do"—all in vain. She was desperate to find some sort of employment even though a friend urged her to wait until she was more recovered.

Sara wrote Mattie that her friend didn't "know how light my pocket is or how much I feel the need of doing something for a livelihood, something that my strength will allow—and that will, I fear, be a difficult matter."

Sara added that JG was off collecting in the Mount Shasta area and loyally proclaimed that he deserved to be paid twice what he'd been offered for a government publication on the locusts that had ravaged the harvest the previous fall. That segued into a passionate political rant in staunch defense of JG and her own hatred of "the solid wicked South

Fig. 15. Mattie Plummer Everett in her early forties, around 1880. Photo by author. Original at the UC and Jepson Herbaria Archives, University of California, Berkeley.

who may thank the North that they were not snuffed out but who now are plotting to yet destroy the Union & get hold enough that which they tried to destroy." She added vehemently, "If I were a man, I surely would stump the country in the coming campaign for Garfield & Arthur. It will be a sorry day for the whole country if the Democrats get into power."

In addition to her own frequent letters home, during JG's trip to Mount Shasta she was keeping up with his correspondence as well. In Asa Gray's Harvard files is a letter in Sara's distinctive handwriting, dated October

11, 1880, but signed "J.G. Lemmon." In it "J.G." writes that "he" will be collecting in Arizona and Mexico for a few weeks before returning to Oakland for the winter's work. It was on this October 1880 trip that JG first explored the Santa Catalina Mountains north of Tucson, discovering two new species of ferns. The finding whetted his appetite to return at some point to southern Arizona—little suspecting that he'd be back the following year as a newlywed.

Ten days later, JG himself wrote George Engelmann from Fort Grant, just east of Tucson, that he wasn't able to afford to go the Santa Rita Mountains. Instead he'd climbed Mount Graham of the Pinaleño Mountains where he found some new species—and saw a native parrot, which, frustratingly, he surmised fed on the seeds he needed to identify the still-disputed Arizona pine (*Pinus arizonica*). He also wrote that he was setting off for Oakland the next day and would then head immediately to Sierra Valley to bring Amila back to Oakland for the winter.

Oddly, JG didn't mention Sara in that letter. Yet, two weeks later, on Sunday, November 7, 1880, she wrote her father from Santa Barbara with huge news: Miss Sara Plummer and Mr. John Gill Lemmon would be marrying later that month!

> Yesterday was a busy day. I am having all my household goods packed in with the dust and rubbish to be sent to Oakland on the Southern on Tues., at 6 a.m. when I go as far as San Luis Obispo about halfway up. . . .
> All my friends here congratulate me warmly on my choice of Mr. Lemmon and say it is just the thing for me as they see the suitableness of the alliance—and they all know him and his sterling worth.

All Sara's household goods didn't amount to much. She figured an auction wasn't worth the effort and that it'd be easier to cull them in Oakland. So her packed belongings included "two chairs, a sofa-bed, a 3/4 bed—good mattress—a pair blankets, quilts, three sheets, pillow cases, a few towels, one cooking stove and a small kerosene stove, a carpet, a few

Fig. 16. Map of the southern Arizona Territory in the 1880s, showing the Santa Catalina, Chiricahua, and Huachuca mountain ranges where Sara and JG botanized between 1881 and 1884. They would return here in 1905 in celebration of their twenty-fifth wedding anniversary. Illustration by author.

odd white dishes, glass jars, knives, forks, spoons &c—all common, but they will be enough and will answer the purpose for ourselves."

And she promised Father there'd be a corner for him—if only he'd come visit.

By this time JG had relocated with Amila (who Sara referred to as "Dear Little Muzzer") and his entire herbarium to Blake House, on the corner of Twelfth and Washington Streets in Oakland.

One week later, on November 13, Sara's last letter as a single woman to her future husband was a mix of the practical—and the sweetly sentimental. On one hand, she wrote, "What a time we shall have in unpacking and casting out and arranging the collection of rubbish!" And on the other she wished he could be with her every hour:

> I pray we may be, O, I devoutly hope we may & then comes the resolve that united we must not overlook anything or neglect little things that will appeal to each individually for the other. Lemmonia, dear, we have assumed deep responsibility & for that reason our happiness will increase in proportion as we show no [illegible word covered by a dried plant, possibly a four-leaved clover] in meeting life as it comes to us. We will try, won't we dear Heart? Then we will shall [sic] get all the happiness we can from our, perhaps, brief existence.
>
> How I wish the preliminaries were done with and that we were settled peacefully in our Herbarium Snuggery.

Miss Sara Plummer and "Professor" John Lemmon were married quietly on Thanksgiving Day, 1880. Their friend and the founder of the First Unitarian Church of Oakland, the Reverend Laurentine Hamilton, performed the ceremony.

Sara, ever proud of her frugality, wrote her brother Charley and his wife, Jean, that she spent only two dollars on her wedding outfit: "a new pair of gloves, $1.25, & for trimming a traveling hat, 75¢" since all the other "fixings" were in her bandbox. (She did point out Mr. Lemmon spent more extensively, around fifty dollars.) Charles and Emily Parry, delighted that their friends had finally made up their minds, attended the ceremony and gave the newlyweds a beautiful gold and silver burnished stand for their dinner table.

All the general and scientific media announced the event, and the *Santa Barbara Weekly Press* noted the "Marriage of Distinguished Scientists," adding "both are scholarly, gifted and write beautifully."

The Civil War might have ended fifteen years earlier, but its echoes still haunted the household. Sara warned her brother that her Lemmonia was particular that people remember the second "m" in his name: The

Figs. 17 and 18. Sara Allen Plummer Lemmon and John Gill Lemmon in photographs made around the time of the Lemmons' wedding on Thanksgiving Day, 1880. Photos by author. Originals at the UC and Jepson Herbaria Archives, University of California, Berkeley.

other branch of the family were Southern Rebels and had changed their name to "Lemon." JG had no desire to be confused with them.

Other reminders of the war moved into the new residence as well:

The old battered sabre hangs upon the wall as quietly as though it had not done execution in battle. He still keeps the old waterproof

hat that he used in those cattle-pens to carry water in to the thirsty and dying comrades, the tin pint dipper used to draw & carry rations in, some of the raw beans served to the poor prisoners and the very fife handed down from soldier to soldier & Mr Lemmon being the last one who didn't succumb, still keeps it as a remembrance of what would sometimes keep the boys' spirits from entirely forsaking them.

You can now understand why he wishes two ms in his name—but enough of this, we are as happy in our snug, cozy home with his dear aged mother, as you, Jean, or any looker-on could wish.

7

"Try to Touch the Heart of Santa Catalina"

Southern Arizona, Spring 1881

→ TEN YEARS HAD PASSED since Miss Sara Plummer established the Santa Barbara Library. Now, thanks to "entering the matrimonial vortex" at age forty-four, she was Mrs. J. G. Lemmon and living three hundred miles away in Oakland. Accounts differ: Perhaps Colonel Hollister bought all fifteen hundred books for $500 and then donated them to the International Order of Odd Fellows. Or perhaps the Odd Fellows bought 2,921 volumes for an undisclosed amount. Either way, Sara was both pleased and relieved, and the books became the foundation of Santa Barbara's first free lending library.

In Oakland the newlyweds settled contentedly into uniting their lives and their botanical collections under one roof. Each day they labored over their immense piles of plants, drying, preserving, labeling and reorganizing specimens, assembling species lists, and sending off sets of rare Pacific ferns to eastern botanists who were willing to pay up to fifty cents per plant. "This, we hope, will keep the wolf from the door," Sara wrote to her brother Charley.

By this time Sara's name was familiar in science circles. In January 1881 John Coulter, editor of the staid *Botanical Gazette*, mentioned the wedding of JG and Sara, "another well-known botanist," adding that the couple "having united fortunes and herbaria, are ready to welcome their friends in their new herbarium rooms."

Thursdays from 1 p.m. to 4 p.m. were set aside for social visits for all three members of the household. Having her nearly eighty-year-old mother-in-law in residence was fine with Sara: She described the still fearless Amila to Mattie as "small & nimble & very bright and is to be with us during the winter. She is excellent company, a great reader. We chum together ever so much."

At Sara's suggestion, the couple had decided to postpone their honeymoon until spring. As JG described it, "My wife, being as enthusiastic and as devoted to botany as I, was the first to propose that, instead of the usual stupid and expensive visit to a watering-place, idling our time in useless saunterings, and listening to silly gossip, we should wait a few weeks, devoting the time to study; then, at the right time, make a grand botanical raid into Arizona, and try to touch the heart of Santa Catalina."

Everyone warned them of the dangers and rigors of the Santa Catalina Mountains, which lay just north of the U.S. Army's Fort Lowell in Tucson. But the more the newlyweds heard, the more determined they were to explore the area. After all, rumors described a high protected valley that even Cyrus Pringle hadn't been able to reach—and, having spent time with Pringle, both Sara and JG envied his "strong frame and good health" and admired his conscientious and painstaking botanical work. Additionally, General Eugene Carr told them no white man—or woman—had ever been there because until recently it was an Apache stronghold.

The Apaches had been enemies of the Mexicans for many years, and the United States inherited that same conflict along with acquiring Arizona and New Mexico at the end of the Mexican-American War in 1846. After several decades of brutal massacres of the Indigenous people, members of all the different bands—five thousand White Mountain, Warm Springs, Chiricahua, San Carlos, Tonto, and Yavapai Apaches—had been forcibly moved or marched to the San Carlos reservation.

So, theoretically, botanizing should be safe. JG wrote that "when we learned all this, the information but intensified the resolution formed on a preliminary excursion the season before that we would penetrate these unknown mountains."

The spring of 1881 was hectic for the Lemmons. On March 7, Sara presented a paper at the California Academy of Sciences, announcing to her family, "I have the honor to be the first lady who has ever addressed that august body."

Fig. 19. Photographer Carleton Watkins's image of Tucson in 1880. Watkins no. 1325, Online Archive of California, Bancroft Library, University of California, Berkeley.

Her paper was a groundbreaking work on Pacific ferns. In it she described fifteen topics in technical detail, including the geography, morphology, and what would later be called the ecology of the plants.

The couple was also busy packing for an overland wilderness expedition, preparing presentations, teaching botany classes, proofreading scientific publications, sending off plant sets, double-checking lists of supplies, arranging rail and wagon transportation, and conducting all the inevitable last-minute correspondence. In addition, the *Pacific Rural Press* had accepted Sara's article for publication, and she and JG finished preparing a technical manual on the ferns, illustrated by Sara. They were also at work on a popular plant guide that they hoped to publish in time for the holidays to sell for a dollar or two each—about twenty-five dollars in 2020 currency. Sara wrote Mattie that she had the manuscript nearly ready and had completed seventy full-color paintings. She'd also used their compound microscope to create illustrations of the spores, root hairs, and more.

The book, however, was never published, and the manuscript as well as the fern illustrations have all since disappeared. She would continue working from the microscope, and other paintings have survived.

At last, after weeks of preparation, the Lemmons were ready to depart for Arizona. Amila agreed to stay home to watch over the herbarium while her "wandering children" explored, and, on March 20, 1881, Sara wrote her father and Mattie:

Tomorrow at 4 P.M. Lemmonia & I start for Arizona and New Mexico on a botanic exploring expedition. The Presidents of Council of the California Academy of Sciences recommended and requested that Gov. [Leland] Stanford give us a pass over their railroads till June 30, 1881. We may be able to renew them if we do well on the trip, i.e. send some good things to the Academy and Gov. S—— in the way of defunct Apache skulls, Aztec pottery, rare plants etc. and it is worthwhile. . . .

I have long desired to see the far-off land of the Apaches and then to go with Mr. Lemmon exploring, and gathering the rare and perhaps new species of flora will be sufficient delight to more than balance the fatigues consequent upon such a trip.

Sara reassured the family they were well provisioned—right down to "the big lunch basket filled with corned beef, crackers, cheese & three or four jars of nice currant jelly, brought in by good, thoughtful friends." How could Mattie and her father not be as excited as she was? After all, she wrote, "the whole outfit and occasion is brimful of interest and enthusiasm, if you will allow the expression."

It was an ambitious undertaking, not one for casual armchair botanists: Sara and JG planned to travel one thousand miles by rail from San Francisco to Tucson, where they'd spend several weeks collecting plants of the Santa Catalina Mountains, from bottom to top. Their railroad passes then allowed them to travel for free to El Paso, Texas, for more collecting.

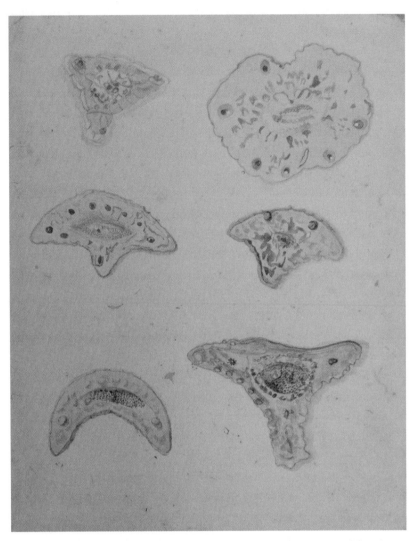

Fig. 20. One of Sara's illustrations made while using a microscope. Sara's handwritten note on the back explains that the sketches are cross-sections of the leaves of six different pines, showing the number and location of the resin ducts. She wrote the trees are "Pinus Lambertiana, P. microphylla, P. ponderosa, P. Bulfourana [*sic*], P. Murryana, and P. Torreyana." Photo by author. Original at the UC and Jepson Herbaria Archives, University of California, Berkeley.

On the return trip, they'd stop again and hire a wagon to head south, probing deeper into the Arizona Territory's unexplored Chiricahua and Huachuca Mountains (see figure 16).

Mattie and Micajah must have felt a twinge of worry as they read: "You will hear from me or us from time to time. With love and goodbye—your good son and new brother Lemmonia and mother join in love, as ever yours, Sara A. P. Lemmon."

Located squarely in the middle of the thirty-thousand-square-mile Gadsden Purchase, Tucson didn't become part of the United States until 1854. By March 1881 it was a lively, thriving town of ten thousand, with four churches, five newspapers, and three schools. The town also claimed to be home to more blacksmiths than any other community of its size in the United States.

Many residents called the climate "agreeable" despite a high of 110 the previous year—and a measurable amount of snow in March 1881. Soon after that last snowfall, and almost precisely a year after the first train to ever reach Tucson had pulled into town, the two middle-aged botanists clambered stiffly down to the platform from a Southern Pacific Railroad car. Despite their frail health, their luggage contained camping gear, walking sticks, sturdy calfskin boots, and more, according to Sara: "a big amount of surplus papers and dryers to change off with & pack plants into. We take several changes of flannel underclothes, the oldest & poorest to be thrown away after they will not hang upon us any longer time—then a supply of tea, sugar, coffee, rice, beans, flour, popcorn, pickles, syrup, canned meats & fruit, butter, cheese &—then ammonia for snakebites, a few medicines for any emergency, each a pair of blankets, pillows & each a shawl for night wraps."

Sara and JG hired a wagon to haul them and their equipment six miles northeast of Fort Lowell to the outer limits of Tucson, where they set up their headquarters in a deserted cabin—deserted, that is, by all except pack rats. They unpacked, sorted their gear, and, too eager to sleep, rose early the next morning to start exploring.

Sara described their outfits and what they carried:

Mr. Lemmon, or *Lemmonia*, as I call him, has a gray suit of strong
duck, boots, the soles of which are covered with heavy gimp tacks,
a big slouch hat, and heavy buckskin gloves.

    Mine is made up from a deep olive-green broadcloth & corduroy
walking suit, presented by Dr. Dunning's wife just before she died
out here. It makes a strong short dress, with turkish trousers met by
leather leggings buttoned and strapped under the heavy calf-skin
shoes, also gimp nailed, long doe-skin gloves and a broad-brim hat
constitute my tout-en-semble.

    Each carry strong wire flower presses to put plants in. Said
presses filled with about 150 folio sheets of thin Manila paper with
felt paper, dryers between every 10 folios for the fieldwork.

The Santa Catalina range is a sky island, rising up from the surrounding
Sonoran Desert, 1.65 billion years of tortured geology, nearly ten thousand
feet tall and stretching seventy miles in length. Looking from where Sara
and JG arrived near what is now midtown Tucson, the shape of the range
is deceptively smooth, its highest rocky crags lurking beyond a gentler
foreslope. Perhaps that's the reason the Indigenous Native American
tribe, the Tohono O'odham, named it Babad Du'ag, or "Frog Mountain,"
honoring the Sonoran Desert toad, a common local amphibian.

At the foot of the mountains the honeymoon couple passed spreading
flats of creosote bushes, evocatively fragrant after rain, and followed
the wandering track of the Rillito River among cottonwood groves and
scattered mesquite bosques. Then and now, as the elevation changes,
so do the vegetation communities, shifting from desert scrub through
grasslands, into oak woodlands that then blend into pine forests. At the
top, seven thousand feet above the desert floor, visitors can be excused
for thinking they've been transported to Vermont when they encounter
scarlet maples in fall or a Canadian winter's snow-covered firs.

The Sonoran Desert, covering central and southern Arizona, is rich
in both fauna and flora and is one of the world's most highly biodiverse

areas. Two thousand plant species are found here, along with 550 kinds of vertebrates and an uncountable number of invertebrates. Currently, about fifteen hundred plant species have been identified just in the Santa Catalina range—but when Sara and JG arrived in 1881, most of those species were either collected but not yet named or still undiscovered.

The couple deliberately planned their trip for spring, one of southern Arizona's two prime times for botany—the second being late summer when the monsoon storms reawaken the parched desert. Sara's artist eye, accustomed to coastal New England and California, was astonished: Green-trunked "palo verde" trees showered the ground with golden blossoms, bright yellow brittle bush still flowered in the shade, the long witches' arms of the ocotillo stretched toward the sky with scarlet fingertips, and barrel cactuses were studded with brilliant orange blooms. Even the wildlife was bizarre. Gambel's quail chittered, their ridiculous topknots bobbling as their call of "*Cuidado! Cuidado!*" warned thumb-sized youngsters of Cooper's hawks swooping through on a speedy and often lethal ambush. One long-tailed bird didn't fly but *ran* past their campsite, and a large pink-and-black beaded lizard lumbered along the wash, its shiny leathery tongue licking up unwary ants. Odder yet, the serpents wore rattles and weren't afraid to shake them in warning.

Creeks, lined with green carpets of sedges and punctuated with hop-bushes, still trickled clear with snow melt, and the descending trill of an occasional canyon wren reverberated off canyon walls. At night coyote choruses woke the Lemmons, alerting them to the scuttle of skunks in search of any escaped food scraps among the dry leaf litter. The magnificent yet improbably comical sentinel saguaros, home to Gila woodpeckers and gilded flickers, towered over it all.

Day after day, Sara and John scrambled up and down steep cliffs and ravines among plants that seemed determined to draw blood: opuntia spines, cat claw thorns, yucca spears, agave bayonets. They dodged the bundles of prickly fallen teddy-bear cholla—while always keeping a watchful eye out for the rumored Apaches. They each lugged water, a little food, plant presses for flowers and leaves, and wads of damp rags to wrap and preserve any prized and fragile ferns they might find.

At first it was all such a contrast to the cool fogs of the Bay Area that Sara paused frequently to inhale the essence of the desert, to savor the sun's heat penetrating that olive-green broadcloth, and to allow her eyes to stretch over the vast expanse. Each evening the sun sank behind the jagged Tucson Mountains, to the southwest hunkered Baboquivari Peak, the Santa Ritas underlined the southern horizon, and dawn broke behind the Rincons. Farther east were the Whetstone Mountains and Tombstone, where the shootout at the OK Corral would occur a few months later.

But spring is short in the desert. As the days wore on, daytime temperatures rose until the heat was "torrid." The closest spring, three-quarters of a mile away, shrank to a dribble so small the honeymoon couple had to squeeze their rubber drinking cups into crevices to get any water at all. They found a tiny cave halfway up the mountain and made numerous trips to ferry all the gear so they'd be closer to the top, their ultimate destination.

The days grew even hotter, the terrain more rugged. And yet, both of them were deeply happy, exclaiming joyously to each other as they found "new glories"—an unknown agave here, a mystery mallow there—and busily spent every evening pressing, packing, and labeling the day's finds. They gathered seeds of one attractive bush with deeply dissected and particularly pungent leaves to plant back at their Oakland herbarium. Later Asa Gray named it *Tagetes lemmonii,* or Lemmon's marigold. Eventually it would become the stock that still supplies nurseries nationwide.

Each day they covered a different route, thinking *this* one would wind its way to the top of the mountain. Each day they'd end up peering down a five-hundred-foot-deep ravine or hopelessly up at an unclimbable cliff. Three times they set out on paths that surely aimed for the summit—and each one dwindled into yet another dead-end in terrain much too rugged to cross.

Finally one afternoon they reached a point high enough they could see another ridge topped by the true upper peaks, invisible from the desert below. Between them and the peaks was "an abyss two thousand feet deep and twice as far across that everywhere separated us from the main mountain. . . . There was no help for it. We must return, baffled," JG wrote. "Beneath us yawned the chasm. Beyond, and far above, stood the guardian pinnacle, between which lay the narrow saddle through which we could not pass."

After two weeks of rugged fieldwork and three blisteringly hot attempts to reach the top, both Sara and JG were exhausted, and they were nearing the end of their supplies. Yet they still weren't completely defeated. They lugged their equipment back down the hot, rocky slopes, and returned to town to rest. They talked to locals, who suggested the other side of the range should be more accessible.

Undaunted, they hired a stagecoach to drive them to the north-facing side of the mountain range, where they borrowed a mule to haul their collecting and camping supplies. They stopped at a mining site called Oracle Camp for several days of botanizing and where Sara, always the diligent correspondent, quickly scrawled a postcard to her family, postmarked April 28. In it she described camp life as "exhilarating when not overworked searching for plants, etc."

They then continued on up the mountain, carrying an introduction to rancher Emerson Oliver Stratton. He was an easterner, a miner, and an entrepreneur who'd established a homestead in 1879, naming it Pandora Ranch because he'd spent all his money and "everything was gone but hope." He described it years later: "When I arrived, I built a good-sized dugout in the side of the mountain, covered it with a dirt roof, and faced it with boards and a door and window I had brought with me. I furnished the place with a stove, a bed, and other items made from boxes. Just outside I pitched a tent for any company that might drop by."

When Sara and JG met them, the Strattons were still living in the dugout, but any kind of available lodgings at all were welcome to the two weary botanists. Stratton described the couple years later to his daughter, Edith Kitt: "When we saw them coming, Dr. Lemmon was riding and she was walking behind. They were very tired and certainly glad to get to the ranch and much surprised to see a cultivated white woman there. And indeed your mother [who hadn't seen a white woman in many months] was glad to see them."

Although he'd never been to the mountain's top, Stratton claimed to know a route that should work. To Sara and JG's enormous gratitude, he offered to accompany them to the 9,157-foot peak—and even better, he could provide both riding and pack animals.

Two days later, the trio set out early, passing abandoned mine shafts and bare hillsides that had been logged earlier. Soon the trail rose so steeply they had to dismount and lead their horses, but eventually, slowly, between pauses to pant and gasp, they reached the highest peak of the Santa Catalinas.

"We christened it Mount Lemmon," wrote Stratton, "in honor of Mrs. Lemmon, who was the first white woman up there. I chopped the bark off a great pine tree on the very top and we all carved our names." (That tree fell during a storm in the early 1960s.)

According to the collection records Sara and JG sent Asa Gray, there at the top was "a fine deer park" with "thousands of acres of pine trees large and abundant enough for fine lumber." It would be another ten years before the Lemmons would recognize the growing loss of American forests and launch their passionate conservation campaign.

JG also reported Stratton shot a huge buck and later watched as a mountain lion carried it away; the lion was so large that neither the antlers nor feet of the deer touched the ground. They saw a new species of parrot—and most exciting for JG, they found the tree he was hoping for, *Pinus arizonica*, a species he was positive was new. (Even today, taxonomic experts still argue about the status of this tree, for now considered to be a subspecies of the Rocky Mountain ponderosa, *Pinus ponderosa* var. *scopulorum*.)

Hardships and all, JG summarized their experiences in the mountain range: "Suffice it to say, perhaps no more vivid and pleasing contrasts, no more new and valuable floral treasures, no more interesting zoological discoveries, can be met with elsewhere in the large Territory of Arizona, than in this terra incognita, this forest in the mountaintops, this museum of natural history, the heart of Santa Catalina."

Perhaps he should have saved his proclamations. Later that same year, they'd find another botanical paradise in a nearby sky island. Then in

1882 they'd encounter a third one, even richer in "new glories." But for now, on May 12, the *Arizona Weekly Star* announced that the Lemmons had discovered many new plants, including twenty-one new species of ferns and a large number of grasses.

Fortunately for the naming of the peak, Emerson Stratton had a valuable family connection: His daughter Edith married George Kitt, whose uncle was George Roskruge, the early Pima County surveyor. Stratton went on to write, "When Mr. Roskruge made a map of the country about 1904, he put in the name Mt. Lemmon."

The plant lists JG sent to Asa Gray and George Engelmann show the couple moving on to El Paso in mid-May, collecting at the top of the Organ Mountains and along the Rio Grande, and even gathering poplar leaves in the streets of Franklin, Texas, now part of College Station in the Brazos Valley.

From El Paso they took the train back west, disembarking in Willcox, Arizona, and traveling by wagon fifteen miles to the Dos Cabezas Mountains, where they collected a few oaks and even a maple similar to one they'd gathered in the Santa Catalinas ("yields good sap for sugar," JG commented in his notes).

They then continued on south another thirty miles toward the Chiricahua Mountains, where they stayed in Teviston (renamed Bowie in 1910) along the railroad line (see figure 16), becoming acquainted with its namesake, Captain James H. Tevis. When they were able to arrange for a wagon to carry them the thirteen miles across the desert to Fort Bowie, they met a surgeon named James Ord, the brother of General Edward Ord of the Western Division, one of Sara's Santa Barbara friends.

The fort itself was near the infamous Apache Pass in an area still vulnerable to Indian attack—but oh so tempting to the couple, since it had never been botanized before.

By now it was the end of May and scorchingly hot, but JG made several forays out from the fort and found a new species of wild potato, a discovery that made them both yearn even more fervently to come back in a couple of months once the summer rains had started.

Dr. Ord assured them of free housing at the fort if they returned, and Colonel George W. Baylor, commander of the Texas Rangers, promised to keep Sara and JG safely under military protection. Since Baylor had been a Confederate, one can imagine some lively discussions between him and JG, the Union veteran and Andersonville survivor.

With these reassurances, the Lemmons packed up their hundreds of plant specimens, caught the train to Tucson, and returned to Oakland. No sooner had they arrived home than they immediately renewed their railroad passes, determined to return to southern Arizona once the summer rains set in and the "flora will be changed and renewed."

8

"An Extreme Outpost of Civilized Life"

Southern Arizona, Fall 1881

→ THE LEMMONS RETURNED TO Oakland from their unusual honeymoon in the Arizona Territory to find, as JG wrote to Dr. Gray, that "Mother has kept the place in good order, is well & hearty. She joins my wife in best wishes & kind regards."

In July Sara wrote her father, saying, "We are very happy in each other and in our harmonious labors and move along quite like people who have been wed a longer time. I think you would like your new son & I am sure he would like you, if you could only meet. Cannot you come out and spend a sunny winter with us??" That visit never happened, and another three years would pass before Micajah finally met his new son-in-law.

Throughout those "harmonious labors" and despite the collection labels that read "J.G. Lemmon & wife," JG was always quick to acknowledge Sara's share of their work. In his letter to Asa, he wrote, "My enthusiastic little botanical help-meet has accompanied me everywhere, climbing peaks, penetrating gorges or roaming plains, so please give her equal credit, and if anything beautiful is found to be new & you approve of the suggestion, please dedicate to her; and if a new *genus* is found, please give it her maiden name of 'Plummer' perhaps in the form of 'Plummeria.'"

That summer Sara proved she could conquer public speaking as well as peaks and gorges. JG had committed to giving a talk to the Chautauqua Literary and Scientific Circle but was still so exhausted from the Arizona trip that Sara agreed to help out. As she told Mattie on July 17, "'Yours truly,' instead of giving one talk, without notes, gave three, two of them over an hour's duration and the last from 2 till 5 P.M. all with Illustrations on the blackboard, using my portable Herbs and the Ferns of the Pacific

Coast & some fresh ones from the woods of the vicinity. There seemed to be quite an interest & what was the most gratifying, a growing interest to the last minute. It was very pleasant to us."

In 1881 the Chautauqua Literary and Scientific Circle expanded to California and held its annual conference in tents among the pines at Pacific Grove Retreat near Monterey—now famous for being the home of John Steinbeck and the overwintering spot for thousands of monarch butterflies. (In 1939 it became a misdemeanor in Pacific Grove to "molest" a butterfly.) The organization had been founded in 1878 near Lake Chautauqua, New York, with two goals for adult education: "to promote habits of reading and study in nature, art, science, and in secular and sacred literature, and to encourage individual study, to open the college world to persons unable to attend higher institution of learning." The Lemmons were enthusiastic supporters and were frequent lecturers during the next few years. As of 2021 the program is still going strong.

At 9:30 the morning of July 2, 1881, the Republican president James Garfield was shot at a train station in Washington DC by Charles Guiteau, a delusional failed politician with an imaginary grudge against the president. Garfield was transported back to the White House and then to the New Jersey shore in hopes that the fresh ocean air would help him recover. All summer the country waited to find out if Vice President Chester Arthur would replace Garfield as president.

In August the Lemmons once again packed up their camping and collecting gear and headed out to the Arizona frontier—despite being delayed by JG's dysentery and then a lingering relic of spinal meningitis that threatened to fell Sara.

This time they planned to make their headquarters in the Chiricahua Mountains ("pronounced *Cheery-cow-a*," explained Sara) first at Fort Bowie, then later at the old abandoned Fort Rucker, forty miles deeper into the mountains. Although the military assured the couple the Apaches were safely corralled at the San Carlos reservation, the Lemmons would be mostly on their own in what was almost wilderness and home to mountain lions, bears, and the occasional wolf. Sara mentioned neither

Indians nor wildlife to Mattie and Micajah, instead writing, "Here we expect to find many new and rare plants to contribute to the scientific world." But she did add, "This time each of us will carry a sheath-knife for convenience and defense. We have never carried even as much as a knife for defense before, but where we are to explore this time, we may need such a thing. At any rate it is always handy for daily use at meal time where one is 'roughing it,' as we shall be obliged to be at times."

JG, the former soldier, commented that weapons not only added unnecessary weight but also were never at hand when needed.

Like the Santa Catalinas, the Chiricahuas are a sky island, this one born of a violent volcano. The range was shoved up out of the caldera's basin 27 million years ago and lies at the junction of the Rocky Mountains and Mexico's Sierra Madre on the north-south axis and the Sonoran and Chihuahuan Deserts on the east-west axis. This collision of geology and life zones results in an astonishing diversity of both flora and fauna—as JG and Sara would soon discover.

On August 24 she wrote her father again, this time on printed stationery from J. H. Tevis, Bowie Station, Southern Pacific Railroad. She apologized that her letter of the previous Sunday was delayed by rain and washouts so severe that no trains were able to get through. After spending two unexpected nights in Los Angeles, they had finally reached Tucson, where they spent another night, then caught the train for Teviston, arriving at 10 p.m.:

> The inhabitants of about a half dozen houses in bed—a saloon was open. Four or five men came to our rescue, as we were landed by the side of the track, all dark except when the lightning flashes revealed the distant mountains across the wide open plains. It was a desolate scene to us tired mortals. Strangers though they are and in league with the cow-boys, we are told, they nevertheless helped us right over the difficulty and escorted us with our various packages to Capt. Tevis's house a few rods away. He came out & welcomed us, giving up his room and sleeping on the open veranda. He was very kind to us when we were here before.

The "cow-boys" Sara mentioned were not benevolent Roy Rogers figures. These men, she explained, were desperate outlaws who stole the best horses and hundreds of cattle from army posts and ranchers alike. They were often as much a foe for the military troops as the Indians.

The next day Captain Tevis helped the couple move to another house owned by the railroad—thus reclaiming his bed. Here they could wait till the telegraph wires were back up and Dr. Ord, the Fort Bowie surgeon, could come retrieve them.

August often brings summer monsoons to southern Arizona, and Sara was struck by how much the landscape had changed in two short months: "Wherever the eye turns on these vast deserts it is one dense mass of verdure and blooming plants. Where here before, nothing but desolation of dust—heat—& cacti seemed to hold the land. Now the grass is fine, and many species may be found together with curious plants. The whole country looks like the East in June."

She was also awed by the beauty and power of the monsoonal storms:

As I write, on all sides are showers. I count six showers over the distant and near mountain peaks, and just here the sun is shining brightly. Every now and then it lightnings & the thunder rattles loud. Cloudbursts are frequent & sometimes come tearing down the mountain ravines, the water & debris several feet high like a big wall of water, roaring and sweeping along the plain for miles, finally spreading out & disappearing over the parched and hungry earth.

Frequently people are suddenly overtaken by them, swept away & drowned, beaten often all to pieces & and buried in the trees, boulders and rubbish.

Fort Bowie was thirteen miles from Teviston, nestled in the eastern Chiricahua foothills where the Lemmons could botanize the wild country while remaining within the reassuring protection of the military.

For centuries, the Chiricahua Apaches had roamed the canyons, grasslands, mesquite bosques, sycamore-lined creeks, and pine-covered mountains throughout southeastern Arizona, southwestern New Mexico, and down into the Sierra Madre mountain range of Mexico. By 1872 the U.S. government had driven the tribe onto a reservation that covered much of the area south of Fort Bowie and north of the international border. But after Custer was defeated at Little Big Horn in 1876, the policy was suddenly revoked, and the Bureau of Indian Services forced the Chiricahua Apaches to move to the hated San Carlos Indian Reservation, which was already overcrowded with other bands of Apaches.

By 1881, when the Lemmons arrived in southern Arizona, food was scarce on the reservation, and malaria was rampant, wiping out half the population. Warriors, led by leaders Victorio, Geronimo, and Juh (pronounced "Ho"), frequently broke out to raid local ranches, killing or kidnapping the occupants.

Fear breeds hatred, as clearly shown in Sara's letter to Mattie on September 5:

We have been deeply anxious here for the past week owing to the Indian troubles. Of course with you the news from so great a distance sounds like a myth, but with the dwellers in the midst, it is a solemn reality. Nothing that prowls over the earth is probably more to be dreaded than the Apache Indians. They are sly, treacherous, revengeful, cruel & love to shed the blood of the white man as much as we take pleasure in killing rattlesnakes—and just as difficult will it prove to subdue that instinct in them as it would be to wean us from the desire to kill a snake. As we feel it to be our natural enemy, so they look upon the white man. Occasionally a rattlesnake, tiger or other animal is caught, caged and partially tamed, given their freedom and how soon they will lapse back to their native ways—and so it is with these savages.

Given the terror that gripped the local white population, the Lemmons didn't care to venture far from the railroad station. Sara added, "We mean to keep on the lookout and not stray off from protection."

Several days after arriving in Teviston, they spotted a six-mule military ambulance headed toward Bowie Station—with occupants wildly waving handkerchiefs in greeting. Realizing Dr. and Mrs. Ord had arrived to escort them to the fort, Sara and JG enthusiastically swung their umbrellas in response.

After a "substantial lunch," they loaded up the wagon and headed to Fort Bowie through Apache Pass, tucked in the foothills that linked the Dos Cabezas Mountains to the Chiricahuas.

Sara and JG settled in easily, as she described a few days later:

This I can say that we are safe and very comfortable in the same building with the Post Surgeon—Dr. Ord. We have four rooms for our moving about in. Two we do not use, only leave doors open for good ventilation. Each room has a fireplace. One we use as a sleeping and living room, another as the Herbarium—and it is nearly full of plants, packages are all around on the floor and under pressure. One table covered with well-preserved specimens in the deep recess of the window with a small table devoted to analysis and study of plants.

I am now sitting before a large round table in the living room before the open door that gives a view of the 1st and 2nd officers' houses on the right, then across the parade ground is an open square plaza, upon which is a cannon, trained to guard the entrance of the Apache Pass.

Our door looks toward the West and post-traders, general quarters hospital—soldiers quarters on the right—this leaves the parade ground in the center, which just at this time looks very attractive as it is with grass and flowers. All around and almost above us are mountain peaks. We are at an elevation of 6000 feet—as I am just told—in the heart of the Chir-ri-ca-hua [sic] Mountains. This is one of the extreme outposts of civilized life. It is wild, picturesque & grand.

Although the Lemmons were safely within the fort's protection, the Apaches were still a worry: "Just beyond this place is the graveyard in which are buried about 60 people, most of them bearing this touching notice on the wooden headboard: 'Killed by the Apaches,' 'Killed by the Indians,' 'Unknown,' or 'Supposed to be——.' We read with heavy hearts their sad fate, then as we left, gathered several species of plants and on the way back about a mile from our quarters picked up a bombshell as an appropriate souvenir of the place."

They spent a couple of days exploring and botanizing near Apache Pass, where plants grew luxuriously, partly due to plentiful summer rains and also because the pass held the only freshwater spring for miles. It had been a longtime traditional camping spot for the Apaches, particularly the band led by the chief Cochise. In 1858 the Butterfield Overland Mail company established a mail station at the spring, which caused some friction with the tribe—but nothing like the hostility that boiled over three years later in the 1861 Bascom Affair. In that event, the hotheaded Lieutenant George Bascom first falsely accused Cochise of kidnapping a young boy and then tricked the chief into coming into the soldiers' camp. Cochise escaped, and in reprisal, Bascom ordered his men to kill Cochise's brother and the five warriors who'd accompanied him.

The following year was the bloody 1862 Battle of Apache Pass, in which white soldiers used howitzers to mow down 150 Apache warriors, led by Cochise and his aging father-in-law, Mangas Coloradas. That battle convinced the army to build Fort Bowie to protect their troops and the surrounding settlers—as well as the water source they now claimed as all theirs.

In January 1863, Mangas Coloradas recognized that fighting the white men was futile, and he reached out for peace by traveling alone to Fort McLane in southwest New Mexico. Under the guise of hosting a peace council, the commanding officer, General Joseph Rodman West, raised a white flag—and then ordered his men to capture, torture, and murder the Apache chief.

The next day the soldiers cut off Mangas Coloradas's head, boiled it, and sent it to Washington DC. Although the head was supposedly shipped to the Smithsonian for study, no record of it exists.

That vile, monstrous crime was even more horrific, according to Asa Daklugie, the son of Chief Juh and nephew of Geronimo, because to an

Apache "the mutilation of the body is much worse than death, because the body must go through eternity in the mutilated condition. . . . That meant that their great chief must go through the Happy Place forever headless."

Not surprisingly, the Apache hostilities continued unabated for the next eighteen years as they fought to protect their culture and ideals. Then U.S. forces killed Nock-ay-det-klinne, another Apache chief, at the Battle of Cibecue Creek, 120 miles north of Fort Bowie. The soldiers had been forced to retreat to Fort Apache, south of what is now Show Low, and rumors flew that the Indians had killed General Carr, his officers, and all the inhabitants of Fort Apache. Commanders from Fort Bowie and Fort Thomas (now Safford) sent reinforcements who discovered that the troops had been able to hold Fort Apache after all—with the exception of eight dead men who had to be buried.

Sara and JG arrived at Fort Bowie August 28, 1881, the same day as the Cibecue Creek battle. Two weeks later she wrote,

> Lieutenant Clark relates a touching incident [that occurred] on their way back to Fort Apache from this mournful errand. In a narrow defile of the mountains a lean bloodshot-eyed and wounded dog came up to them & and by its actions led them aside from the trail, and to their horror was a man decomposing who had been killed, then following the dog a little farther on, they found another man and this dog had actually worn down a path going from one to the other, itself wounded in the shoulder & starving, kept guard of these two dead friends. Lieut. Clark says, "no words could express the relief & joy that dog showed as he knew his faithfulness had been rewarded." He said that the dog must've been there a week at least. That dog, better than an Apache, was taken to the Fort and Gen. Carr now has him. He is being well cared for.

Fired up by the Cibecue Creek battle, a band of Apaches led by Juh, Naiche (youngest son of Cochise), and Geronimo bolted from the Fort

Apache reservation. Rumors had them riding south directly toward Fort Bowie, determined to avenge their dead.

Each snap of a twig made Sara and JG jump as they surveyed the woods near the fort, gathering unfamiliar species, sometimes crawling on hands and knees to examine every tiny plant.

One day a tall, yellow, daisy-like flower caught Sara's eye. She didn't recognize it, so, hoping it was a new species, she carefully collected it and folded it in half to fit on the absorbent paper. She gathered a second specimen of the same plant, short enough to fit without bending, and then delicately placed both in the press, strapped the two boards tightly closed, and added it to their growing pile.

There's little doubt that Sara was as much a field scientist as JG by now—in fact maybe more, as her field notes are much more decipherable and complete than his.

Later that month, she wrote her family, "After a few days when the rainy season is over we hope to go on to a deserted camp, Fort Rucker, 40 miles distant where the flora is said to be very fine. But more of this later—"

9

"Eleven Days of Dungeon Life"

Southern Arizona, Fall 1881

↬ AFTER TWO WEEKS OF productive botanizing near Fort Bowie, and then a delay caused by colds and "inflammation of the bowels," on September 19, Sara and JG set off to Fort Rucker with a driver and two guards in a military ambulance drawn by four mules. It was the same day that President James Garfield died, seventy-nine days after being shot. Ironically, it wasn't the bullet of Charles Guiteau that killed him—it was the infection carried into his body by his surgeons' unwashed hands. As Chester Arthur had been vice president, he took over presidential duties.

Sara and JG, probably unaware of the national news, were more concerned about the two days of hard travel, rattling through the Sulphur Springs grasslands into the oak woodland canyon formed by the sycamore-lined White River, and finally emerging into the natural amphitheater of Rucker Valley, home to what was by then Fort Rucker (see figure 16). Originally named Camp Supply when it was established in 1878, then renamed Camp Powers, the fort initially lay on the southwest edge of the mountain range, next to the river.

The camp was moved upstream and renamed a second time in 1879 in honor of a young soldier, John Anthony Rucker, who drowned when he rode into the flooded creek in a supposed attempt to rescue Austin Henely, his West Point buddy. John Rope, a White Mountain Apache scout, provided a different account. He reported that he, a couple of military suppliers, and the two young soldiers had waited out the storm in the fort's saloon. Once the rain stopped, he and the two civilians rode across the river safely, one by one. Then Rucker and Henely, fortified by their session in the saloon, plunged impetuously into the river side by side. The raging current knocked the two horses together, and both men fell off their mounts and drowned.

By 1880 Apache activity had shifted away from Fort Rucker, and the post was closed. A few mostly roofless adobe buildings still surrounded a broad, level meadow, an ideal drilling ground for the cavalry.

In October 1881 Sara and JG bore yet another letter of introduction to the valley's only remaining resident, Robert Monroe, who lived a mile or two from the former fort. Later known as the Hermit of the Chiricahuas, he was a former New Yorker, according to some, though others detected a southern accent and said he was originally from Virginia.

Either way, he wasn't fond of company.

Mary Kidder Rak, who bought the camp some sixty years later, wrote that Monroe had built a *jacal*, a corral made of "juniper pickets, chinked with mud and rocks and roofed with a thick thatch of bear grass." Here he lived while the fort was functional, mostly off the land but with occasional trips to the supply store. After several months, he'd expanded the cabin, reinforced the door and walls, and then began buying powder and fuses at the store, saying he'd begun doing a little prospecting.

People considered him somewhat odd but thought it wise to avoid asking too many questions. Besides, he did have some medical expertise and was willing to set broken bones and remove bullets without charging a fee. Perhaps that's why he was known as Doctor Monroe.

The only surviving records of the Lemmons' time in Rucker are several drafts of "Botanizing in Apacheland," an article written later by JG. Here he described their host:

We bore a letter addressed to Dr. Robert Monroe, the only resident of the valley—and said to inhabit a retreat three miles up the river—and after a long search what a queer little old specimen of a hermit we found him to be!

Dressed in cast-off soldier-clothes, a world too large, a tattered straw hat poised on one side of his head, he peered out of a little hut, his hand stoutly grasping the door-hook, his small grey eyes flashing from face to face, his long aquiline nose almost meeting his up-turned chin across a toothless mouth, the lower jaw and his lips in constant motion—these uncanny features however were disregarded the moment he recovered from surprise and found speech: for his language tokened a keen intellect, and his manner a good knowledge of the world.

Despite his reputation for being a hermit, Monroe was so delighted with Sara, JG, and their scientific pursuits that he immediately insisted they stay with him in his cabin instead of camping at the abandoned fort. They accepted his offer.

Their visit started out peacefully. The Lemmons explored and collected in the autumnal daylight and at sunset moved into the primitive cabin nestled against a ridge with the sounds of the river as a nightly lullaby.

Even though the two scientists regretted not catching up on notes and labeling specimens every evening, they delighted in the hermit's company. Dr. Monroe was such an exuberant storyteller both Sara and JG were sorry they lacked the stenographic skills to record the tales that came fast and furious. He in turn relished Sara's cooking: Weasels had recently decimated his garden except for a few vegetables, and he was especially partial to Sara's squash fries—his medical skills hadn't included preserving any of his teeth.

One evening, Monroe hinted at something remarkable he might show them. After a few more days, he decided they were trustworthy enough to share his pride and joy: a tunnel he'd dug that passed all the way through the ridge behind the cabin where they sat. That evening he lifted a heavy curtain that hung in the back wall, as JG recounted later:

A low narrow opening, emitting stifling odors, led to the dark interior. Following the interrupted lantern light, we heard a sepulchral voice, warning: "Bow well the head, Stoop very low, So you may hope to farther go . . ."—but hardly in time to prevent bumping against boulders half-projecting from the roof while we were stubbing our toes against similar obstructions in the floor. "Left to break the necks of intruders," explained the miner.

On we bumped and stumbled for about 25 feet when we arrived at a cul-de-sac seemingly the terminus of the tunnel, but while we were wondering and about to ask, "What's next?" the hermit fumbled a moment among some roots on one side—when the wall gave way,—it being a barrel-stone door supported by a concealed post, opening to an elbow "to shut out the light, and confuse enemies."

The tunnel continued another forty feet beyond the zig-zag where the hermit had dug out another larger elbow "for occupation when Apaches were expected." Here he had stored "a bundle of old musty blankets, a box of very stale domestic bread, a barrel of water and the hermit's assortment of rusty guns and bayonets." The tunnel then wound its way to the far side of the ridge where it opened into another small shack and the corral occupied by his chickens and two burros.

Mary Rak confirmed JG's description when she wrote in 1945: "The completed tunnel, still intact, is approximately ninety feet long from mouth to mouth, about four feet wide, and a six-foot man may walk through it without stooping. Moreover, it does not go straight through the hill, but midway has an offset in the form of the letter Z, so that a bullet fired from either mouth must hit the wall. A lone man may there stand off an attack, no matter from which end it is made."

Dr. Monroe proudly showed off his accomplishment to the Lemmons—and then shared the pièce de résistance:

As standing erect now, the wizard of the dungeon looking very wise and confiding, began unfolding the climax of his promised revelations. Stooping, he maneuvered the ends of two protected trenches that he had excavated along the floors of the tunnels (unnoticed by us until then) in which were placed lines of miner's fuse leading from the central elbow each way to the cabins and into a half-keg of gun-powder, covered by a bushel of small cobble-stones concealed under the floors!

The eyes of the inventor gleamed wickedly (we recalled the expression later) as he disclosed this device and declared that by the simple possession of one match he "could blow the whole Apache Nation into shoe-strings and jumbled bones!"

Once back outside the tunnel—politely, quietly, and just between themselves—Sara and JG laughed, if nervously, about the "demented" miner and his ridiculous plan.

As the days went on, Dr. Monroe told fewer stories and grew increasingly morose and moody—until one evening, for no apparent reason, his temper exploded. He suddenly leapt to his feet, grabbed his rifle, and threatened them, shouting, "Mr. Lemmon! You are lying to me! The officers at Bowie have sent you over here to squat upon my premises and rob me of my ranch! Before I'll let you do that, I'll shoot you both!"

Perhaps Monroe forgot he was dealing with a man who'd survived three years of Civil War conflict and both the Andersonville and Florence prisons:

> The fire of insanity glittered in his eyes, his chin rose to his nose and dropped in quick succession, and he began bringing his gun to bear upon me when I sprang to my feet, thrust my hand behind me and yelled, "Dr. Monroe, Sit right down there and shut your mouth! or I'll blow the top of your crazy old head off!"
>
> Before my wife had time to do more than spring between us with blanched face and out-stretched arms, the demon of insanity was exorcised; his face underwent instant change, he looked aghast, languished slightly, dropped into a seat for a moment, then rose with an apologetic bow, hung up his rifle—and resumed his usual avocations.
>
> Thereafter, during our stay, prolonged as it was by several days—the demented Hermit of the Chirricahuas [sic] was as agreeable and considerate as could be desired.

One afternoon several days later, Sara was alone by the cabin catching up on field sketches. JG was off finding more plants and the doctor was out hunting, when a courier galloped into camp with a message from

Fort Bowie's commander. The soldier didn't even dismount as he quickly handed the note to Sara and cantered back down the canyon.

The news was grim: The Apache chief Juh, accompanied by a band of warriors, had escaped once again from the San Carlos reservation and was headed their way. "'I am powerless to rescue you now,' wrote the Major. 'Our whole force—joined with that of other garrisons—is engaged in picketing the towns and patroling the roads. Will send for you as soon as possible. Hope you are not far from Dr. Monroe's cabin.'"

Suddenly the tunnel no longer seemed ridiculous. Sara abandoned her watercolors and sprang into action. Within a few hours, JG and Dr. Monroe returned to the cabin to find

> Amabilis with quiet mien and determined face was resting after her labors. She had re-filled the water-barrel, carried all provisions and bedding into the elbow, and had just finished barricading the cabin-door with every moveable article in the domicile.
>
> Ah! That secret Refuge Tunnel! How glad were we *now* that we knew of its existence! how ashamed that we had ever spoken of it, slightingly, to each other, or doubted the sanity of the miner who excavated the long channel. Yes, and we blessed the conspirator who devised the fuse and gun-powder plot!

Ten days later they were still hiding out in the tunnel, and the Lemmons' relief had long since worn off. They were sick of being cramped up in the dank, cold, dark dungeon, and the once-soothing lullaby of river flowing over the rocks now sounded like approaching human voices and terrified them several times a day. Time dragged as day and night blended into dark gloomy fog. They often shivered for they dared not reveal their presence by lighting a fire—nor could they spare any candles.

One day they heard hoof beats outside and prepared for an attack. JG seized a musket, Sara flourished a bayonet, and the hermit gleefully grabbed the matchbox. JG described what happened next: "Continued attention developed that the first steps were those of a single animal, and the sound began to recede. We crept out to the cabin and peered out of the window to discover a white man passing by."

Relieved, the trio ran out to talk to the man, a stockman hurrying home to New Mexico. He told them Juh's warriors had killed and burned their way to the nearby Whetstone Mountains, but the cavalry blocked them from Rucker Valley and were chasing them south toward Mexico.

Finally, after eleven days, clouds of dust rose at the far end of the valley. Sara trembled as she wondered if the horsemen were a crowd of angry, bloodthirsty Apaches. But no—a glimpse of army blue flashed through the oak trunks, and all three of them thought they'd never seen anything more beautiful than the U.S. Cavalry as the troops trotted onto the grounds of Camp Rucker.

One soldier handed them a welcome message from Major Rafferty. After two days' rest, the cavalry would take the Lemmons back to the safety of Fort Bowie. Sara wept, silently, in relief and joy; JG merely bowed his head.

Dr. Monroe declared loudly to anyone who'd listen, "I wasn't a bit afraid at any time!"

Eleven days of being crammed into the dark, dank tunnel, not knowing if they'd be killed by Apaches or by the demented miner, had taken its toll on the Lemmons. But JG was still determined to find the elusive new cypress he'd been hunting for, and he set off early the next morning. He spent all day clambering around the slopes above the cabin despite a wild thunderstorm and was worn out even before he forded the flooded creek. Exhausted, he returned to camp long after dark—triumphantly bearing the bark and cones of the cypress.

The next day Sara and JG bid a relieved farewell to Monroe and rode out of the canyon as far as a local ranch in a miserably uncomfortable springless government military wagon where JG, ill and worn out from his exertions, lay stretched out flat on top of hard ration boxes. Once the couple arrived back at Fort Bowie, Dr. Ord examined JG, diagnosed "nervous fever," and warned Sara of the possibility of typhoid pneumonia.

Sara immediately arranged for them to start for Oakland as soon as her husband was strong enough.

By October 19, JG was still so weak he couldn't sit up and had to dictate their botanical correspondence to Sara to go along with all the specimens she packed and mailed. He gradually regained enough strength for rail travel, and they finally made it home to Oakland on October 23. Sara described JG's condition to her family:

> Lemmonia just recovering from quite a severe attack of nervous fever. He is yet very weak. So that his head swims as he walks about, and how thin he is to be sure! But we hope now that he is home and our good dear mother's watchful eye is over him, that he will soon be as well as ever.
>
> The trip was a very severe one upon both of us, owing to the added burden of nervous anxiety about the Indians, who put us in real danger for several days, but I think I wrote to you about our experiences in a long letter while at the Hermit's.

Sara mentioned that letter earlier as one that went with the military escort to Dos Cabezas the day they left Rucker Canyon. Sadly, it disappeared, taking with it Sara's account of their time in the tunnel.

10

"Happy in Our Work & in Each Other"

Oakland, 1881–82

↣ IN ADDITION TO JG'S illness, Sara came home injured as well. She'd been walking through Fort Bowie's commissary department when her foot slipped through a hole in the floor.

"It almost broke my leg," she reported to her family. "The knee is sprained and the shinbone very sore—the skin broken in two places."

In November, three weeks after returning to Oakland, Sara and JG still struggled to recover from their ordeal. Every night, dreams of approaching Apaches jerked Sara awake, bolt upright and gasping. She told Mattie, "We have both been so entirely fatigued and ill from our last trip into Arizona as to hardly summon any of the usual wide-awake grip on every day affairs."

JG was still battling his nervous fever when dysentery struck him so severely he couldn't sit up—just as "an inflammation of the bowels" attacked Sara. In addition, her injured leg wasn't healing: "I tried to ignore it, but it wouldn't be passed by, and the trouble has seemed to increase so that it was so swollen and grew so lame that I have been obliged to devote more than a week of close attention to it. By bathing it in Tincture of Arnica after a hot salt water fomenting have rapidly been overcoming the difficulty, but have been obliged for 2 weeks to remain within doors & very quiet."

Her next strategy was to use "St. Jacob's Oil," a heavily advertised patent medicine—in which the active ingredients were chloroform and turpentine. Four months later, though, her leg still bothered her.

Being home in the Lemmon Herbarium, however, wasn't all bad: Sara reported they'd made "a fine capture of rare plants, including some beautiful ferns." They'd been busy sorting and labeling and mailing, and hoped to do well, fiscally, with their "splendid" collection.

"But time will tell & further in the season I hope to tell you of the results. Not very lucrative to be sure, but along with the meager purse comes some honor and not a little contribution to the world of science," she wrote.

That world of science showed up frequently at their doorstep: Edward Greene, who had just taken a position as head of the herbarium at the California Academy of Sciences, came to admire the Arizona collection. Charles and Emily Parry were weekly visitors: "We like them much," said Sara. On December 5, Sara wrote George Engelmann to confirm a shipment and mentioned that they and the Parrys had recently visited their friend John Muir.

Muir had moved to the Bay Area in 1875, and, like the Lemmons, he and Louisa (Louie) Strentzel had just been married in 1880. The Lemmons and Muirs socialized together, and Sara described the naturalist to her sister, saying he wrote for *Scribner's Monthly* and "he is a genius and very genial—a good friend of ours."

The Muirs had moved to Louie's father's fruit farm in Martinez on the East Bay of San Francisco. After years of odd-jobbing, not to mention a thousand-mile walk from Louisville to Florida, John Muir wasn't easily domesticated. Newly wed or not, he still had the travel itch. Before he gained fame as a naturalist, writer, and conservationist, he was a botanist and glacier geologist and was the first to propose that glaciers were living beings that advanced, then retreated while, very slowly, carving out gigantic valleys. (Although Muir was regarded worshipfully for decades, recent research has revealed his less admirable beliefs that smacked of both colonialism and racism.)

In 1879 a whaling ship named the *Jeannette* had been lost in the Arctic off Point Barrow. Two years later, a springtime search expedition was mounted using a steamer, the *Thomas Corwin*, to try to find her. Muir happened to know the captain and jumped at the opportunity to explore the northern reaches of the Bering Sea and its glaciers. His journals and sketches of the desolately spectacular land- and waterscape they traveled for six months and fourteen thousand miles were published, first as a series of letters to the *San Francisco Evening Bulletin* and then again, many years later, in 1917, as *The Cruise of the* Corwin: *Journal of the Arctic Expedition of 1881 in Search of De Long and the* Jeannette.

By September 1881 nasty early winter weather and a broken rudder chain convinced the captain to delay further exploration until another year. The *Corwin* returned home in late October, and Muir immediately sent a collection of plants to his friend and colleague Asa Gray, asking for identification. According to William Frederic Badè, the editor of *The Cruise of the* Corwin, Gray was delighted to find a new species, a rare and delicate member of the daisy family, among the bunch. Gray named it *Erigeron muirii*, partly because ten years earlier he'd written Muir, "Pray, find a new genus, or at least a new species, that I may have the satisfaction of embalming your name, not in glacier ice, but in spicy wild perfume."

Muir's trip was the talk of the botanical world, so Sara and JG were of course delighted to receive a special invitation to see his Arctic plants. Sara, ever her farmer father's daughter, was especially intrigued by the agricultural aspect of the Muir family's "immense ranch": "He has a celebrated vineyard. We went through the acres of vines, many still hanging full of delicious grapes and in great variety, among them the grapes both white & black from which the Zante currants of commerce are prepared." Now called Corinth or Zante currants or champagne grapes, this cultivar is among the oldest form of raisins and is in the *Vitis* genus, unlike red and black currants that are *Ribes*.

From the vineyard the visitors moved on to the "Esquimeaux curiosities" from the *Corwin* expedition. Muir had brought home about 150 species of plants, along with carvings from fossil elephants, and dolls, belt buckles, and fishing hooks made from seal bones.

Sara was most impressed by the seal hunter's seagoing craft Muir had brought back:

a boat, made basket-work-style of light, strong wood and covered with seal-hide about 15 ft long, and in the middle, only about 1½ ft wide, tapering at a point at both ends, all covered with seal-hide, except a hole in this widest part where the native pushes himself in feet-foremost, then sits flat in the boat. Then a fitted piece of the same tough hide is fastened ingeniously about the hole and bought up over the body and thrown closely around the neck. When all is complete, two other natives give him a send off or primitive launch from some icy crag, as the breakers come in, and this is the usual venture of the native fishermen & seal hunters. They tow their game to shore.

The entire day was altogether delightful, ending so late the Muirs ordered a carriage to take Sara and JG home, each bearing an armload of grapes.

As much as she relished the visit, Sara was unimpressed with Muir's plants, as she wrote Engelmann: "Of course everything looked stunted to scraps as compared with the tropical fullness of Arizona flora."

By now Chester Arthur had been president for several months, and the trial for the obviously mentally unstable assassin Charles Guiteau was in full circus mode. Sara's comment to Mattie was, "We follow the trial of Guiteau with interest and disgust comingled. He seems to be overdoing the insanity business."

Thanksgiving brought their first wedding anniversary, and Sara, JG, and Amila planned to spend it peacefully "munching our turkey (i.e. beef soup) in the quiet of our home board." But, as so often happened, illness changed their plans:

> The day before I was taken with chills and quite feverish, ill all day—had invitations out to dinner but could hardly hold my head up. Am up today, and that is about all. Captain Plummer's wife was in the day before Thanksgiving and says the Capt is to be home on the coming Sunday and invites us all over to a turkey dinner. So before and after the honored day perhaps I at least can strike an agreeable balance. It was very tantalizing yesterday—two elaborate dinners of turkey, roast beef, mince pie, etc. were sent in, & not a mouthful could I eat.
>
> However, I threaten to do the subject justice when I get about again.

Part of "getting about" was being swamped with talks to various groups. Five years earlier, in 1876, Dr. Adrian Ebell had founded the Oakland chapter of the International Academy for the Advancement of Women. The organization's flagship program was a study-abroad curriculum, but during its European trip the following year, Ebell died suddenly at a youthful thirty-eight. The academy was renamed the Ebell Literary and Scientific Society in his honor and became a center for learning, as well as civic and suffragist activism: Poet and author of "The Battle Hymn of the Republic" Julia Ward Howe and social reformer Susan B. Anthony were among the visiting speakers. Men were permitted to join, and Sara and JG were both members, as well as frequent and popular speakers.

By now they were an accomplished presentation team, and on December 14 their talk to the to the Ebell Society was titled "Botanizing in the Land of the Apaches": "L & I both occupied the platform, and by turns relieved each other without method or notes. We took up many specimens of plants particular to the region and described them and their habitat. This seemed to hold the audience. . . . Beautiful vases and beautiful flowers adorned the room. Large ferns and Calla lilies with a profusion of rose buds & pinks filled the air with fresh perfume."

The talk was followed by an informal reception when the audience could examine actual Arizona specimens next to a collection from the Royal Gardens of Kew, sent to the Lemmons by none other than Sir John Hooker. Sara told Mattie the newspapers declared the event a "fine success . . . but O dear! It brought no compensation. . . . You know that in time one can starve on honors."

At the time the Biology Section of the Ebell Society was run by Dr. Chloe Annette Buckel, one of the Ebell founders and Alameda County's first female doctor. Born in Warsaw, New York, in 1833, she'd earned her MD from the Women's Medical College of Pennsylvania in two years, then served as a physician in the Civil War. She moved to San Francisco in 1877 for health reasons, but whatever those were didn't stop her from both working for the Pacific Dispensary for Women and Children and founding a chapter of the Agassiz Society to encourage children to study nature. She also established a pure milk commission whose role was to ensure that cows with tuberculosis couldn't supply milk to local dairies.

Annette Buckel apparently recognized a fellow dynamo and activist-in-the-making in Sara. First she asked Sara to take over the Biology Section. Then when the society members asked Sara to teach a "Ferns of the Pacific Coast" class, Annette signed up.

Sadly for the Lemmon budget, Ebell Society bylaws dictated that all classes be free. Fortunately, half a dozen of the twenty prospective pupils said they'd pay twenty-five cents for each one-hour weekly class of instruction for three months.

"This is better than nothing," was Sara's comment.

As anyone who teaches can attest, the classes were just as labor-intensive for the instructors as for the students: "Dr. Buckel has had charge of the Section in Biology of the Ebell Soc and this winter she, with the entire section are listening to a course of conversations and Illustrations upon the Ferns of the Pacific coast, given by me in our rooms. We shall be through in about a month. Lemmonia assists me by the compound microscope to illustrate protoplasm, cellular tissue, stomata, fibrovascular bundles etc., and I in turn assist his lectures by the microscopic illustrations enlarged on the board in colored crayons."

Money was indeed tight that year. Sara and JG's monthly living expenses included rent at $14 and coal and lights for $10. JG's pension was only $4, and on a good month, they could sometimes bring in $30 by selling plant specimens and books. Sara put together a book of plants that was bound for Russia—and received $10 for it. Another set of one hundred labeled Arizona plants went to the Museum of South Kensington, England—for $10.

"So we manage to keep the wolf from howling at our door, although we should hear some growling were we not very economical & simple in our habits," she wrote.

JG was as worried as Sara about money. He too taught a botany class, an evening session for both ladies and gentlemen. His class notes may well have been the basis for a publication the following year in the *Pacific Rural Press*, titled "Hints for Botanical Collection."

On December 20 he wrote George Engelmann: "We are terribly over-worked organizing sets to sell for little money. . . . We must find some other work to do keeping still a firm hold on botany."

"Perhaps," wrote Sara, "in time, like the McCawbers in Dickens, 'Some-thing will turn up.'"

The new year of 1882 may not have brought money, but it did bring some exciting news. Asa Gray had indeed honored Sara by naming Fort Bowie's tall, handsome, daisy-like flower *Plummera floribunda* in her honor. She was so thrilled to have an entire genus named for her, that she described her reaction when she wrote home to Micajah that Sunday: "No person can ever have but *one* genus, as Mattie can explain. I was so delighted with the honor—and such a fine plant—2 feet+ high, that I danced around our big herbarium, overturned the chairs, embraced L & Mother in the most enthusiastic way, and they joined me in the celebration, and declared that it was only right to have something cheery after so many hardships."

To this day, the very specimen that Sara collected in the shadow of Apache Pass rests in the U.S. National Herbarium, part of the Smithsonian Institution's National Museum of Natural History. Sadly, the plant no longer carries her name. Genetic studies revealed that it's not its own genus but is part of the *Hymenoxys* genus. Such is the way of botanical nomenclature.

At the same time, Gray also named a candyleaf *Stevia plummerae* for Sara. She wrote Micajah, "So you see, father, how your name is perpet-uated and honored."

In all, Sara would eventually have several species named for her:

Allium plummerae Wats, Plummer's onion or Tanner's Cañon onion, is a member of the garlic and onion family.

Baccharis plummerae (including two subspecies), or Plummer's baccharis, is the white-flowering shrub Sara found in the hills above Santa Barbara in 1876 and is the first species named for her.

Calochortus plummerae, Plummer's mariposa lily, is native to Southern California. "Mariposa" is "butterfly" in Spanish, and indeed the delicate lavender and yellow petals are reminiscent of butterfly wings.

Ipomoea plummerae, Plummer's morning glory, is sometimes called Huachuca Mountain morning glory to honor the location where the Lemmons first found it.

Lomatium plummerae, Plummer's lomatium, sometimes called glaucous desert parsley, is edible. A member of the carrot family, it's easily recognized by its sticky lacelike leaves and yellow flowers, shaped like tiny pom-poms.

Stevia plummerae, or Plummer's candyleaf, is found only in the mountain canyons of Arizona and New Mexico where it flowers in late summer.

Woodsia plummerae, or Plummer's cliff fern, was one of the plants the Lemmons found while botanizing near Fort Bowie in the fall of 1881.

Another candyleaf, *Stevia amabilis,* was likely named for Sara, but only those familiar with the Lemmons' pet names for one another would recognize that connection. Years later the plant was reclassified as the less romantic *Stevia viscid* (the species name means "sticky").

And what of the cypress JG worked so hard to collect? It's still listed in the World Checklist of Selected Plant Families as *Cupressus arizonica* var. *bonita* Lemmon, and is a synonym for *Cupressus arizonica* Greene. The pine he'd hoped was a new species turned out to be one already known. But he was sufficiently pleased with their efforts that he wrote, "It must be conceded that, after all, the Pleasures have far out-classed the Perils of Botanizing in Apache-Land."

Eventually, the Lemmons would discover 3 percent of Arizona's plants, and dozens of those would carry the Lemmon name. But which Lemmon? Asa Gray clarified that issue two years later when he wrote in the *Synoptical Flora of North America,* about Sara and the genus *Plummera* sp.: "Sara Plummer, now Mrs. J.G. Lemmon, [is] the discoverer. She and her husband have shared together the toils, privations, and dangers of arduous explorations in the wilds of Arizona and California, as well as in the delights of very numerous discoveries: *so that whenever the name of Lemmon is cited for Arizonian plants, it in fact refers to this pair of most enthusiastic botanists"* (author's emphasis).

Other news related to their Arizona adventure was the fate of the Hermit of the Chiricahuas. Sara wrote the family: "Dr. and Mrs. Ord have recently written to us that he went to Tombstone, A.T., 60 miles for his mail & provisions for winter and while away, someone broke into his cabin and stole everything he had including his one suit of good clothes. How sorry we feel for him!"

Various other accounts show Doc Monroe abandoning his cabin in Camp Rucker.

Sara went on in her letter to describe the Apaches in a manner that was typical of the rhetoric of nineteenth-century white settlers, words that today would be widely considered politically incorrect, xenophobic, and even genocidal. It helps to read her words within the context of the times—while remembering both the rights of the Apaches and the terror of the new westerners:

> We only fear that in some unguarded time Monroe will be killed either by stragglers or the horrid Apaches, who by the way are now in Mexico suing for peace with Gen. Terrezes [*sic*—actually Gen. Joaquin Terrazas]. They are just over the American line in Chihuahua State and no doubt will raid upon the borders in the Chiricahua Mountains their old home and a summer haunt. Drat them! I wish they might be exterminated. They are as worthless material in the great economy of civilization as so many man-eating tigers.
>
> How strange our experiences in that border land last season! It seems like a dream, sometimes quite startling!

The early part of 1882 was a season of loss for the Lemmons. First came the death of Sara's friend Dr. Horatio Alger, a Unitarian minister and the father of Horatio Alger Jr., the author of the series of rags-to-riches young adult books for boys. Years earlier, as president of the Natick (Massachusetts) Society of Natural History, Alger had invited Sara to be an honorary member. As a loyal New Englander, she continued to send the society specimens of Arizona ferns that "represent much toil & hardships & dangers. Many of them are rare and bring us $1 each."

In addition, February brought more mournful news: "How sad, sad to me is the telegram just announcing the death of my good pastor and friend, Dr. Bellows! . . . I can hardly realize that he is no more—68 only." Bellows had been her much-loved Unitarian minister, friend, and mentor and had donated two hundred books to Sara's Santa Barbara library.

Another friend, a Mr. Thomas Harris, was so discouraged by the death of his wife and his own inability to find work that he shot himself "just back of the right ear and lived only about five minutes," wrote Sara, never one to shy away from details.

A month later came the death of "the sweet singer, poet, Longfellow! How the lights flicker and vanish!"

And then on Easter Sunday, came a devastating loss:

We are so disturbed and shocked by the sudden death of our minister Dr. Laurentine Hamilton last Sunday that I could not write. Dr. Hamilton stopped suddenly in the midst of one of his best sermons—put his hand to his forehead, bowed his head over on the open Bible, fell backwards, and expired almost instantly. An autopsy showed a sudden rupture of a blood vessel at the base of the brain, and so there was no help for him. This coast has lost one of the best of men, endowed with a most brilliant intellect, a heart large and sympathetic, gave him a hold upon the head & heart of all who came towards him. He was the minister who married us and has always been a warm genial friend.

As if all those departures weren't enough, on April 27, she wrote: "We hear of the death of Ralph Waldo Emerson today. Only a short time ago Longfellow, and a few days since Darwin. The brilliant lights seem to be going out."

At least 1882 brought one new friendship, one that would last many years. Most likely, it was a foggy day in Oakland when Miss Adelia Gates crossed the intersection of Twelfth and Washington Streets and approached the Lemmon Herbarium.

Born in Otego, New York, in 1825, Adelia Sarah Gates began to paint at age fifty and was considered an accomplished decorative flower painter by the time she was fifty-seven. (She is still well enough regarded that six hundred of her paintings remain in the Smithsonian Institution Archives.) By 1882 she'd spent two years at Antioch College, worked as a teacher and governess, traveled alone to Scandinavia, Mallorca, Italy, and Algeria to paint, and lived in Switzerland while studying with the famous watercolorist Emilie Vouga. One of the hallmarks of Vouga's work was using dark paper to depict light-colored flowers, a technique that Adelia would adopt—and eventually pass on to Sara.

But now Adelia itched to learn the science behind her flower paintings—and who better to learn from than Professor and Mrs. John Gill Lemmon? After all, they were renowned western botanists, and Sara's scientific illustrations were so admired that she was the first woman allowed to speak to the august California Academy of Science.

Adelia's meeting with Sara and JG was both congenial and collaborative, as Sara described to Mattie:

> With all my other busy work I am taking lessons in watercolors of a Miss Gates who . . . came to see us—said that if we would exchange and give her lessons in botany she would be glad to assist me all she could in flower painting and sketch from nature. We all see that it will be an added power in our botanic work. Already it is beginning to tell in L's and my lectures. I get more freedom with colored chalk on the boards as illustrations before the classes. You know how it is: one thing helps another—

In spite of all the 1882 losses, Sara told Mattie even though their life "might be denominated 'nip & tuck' with us, we keep on trying and are happy in our work & in each other. Our tastes harmonize, and L is very refined and kind, thoroughly devoted to his aged mother." JG still tired easily, but "he whistles to keep his courage up and often at twilight takes down the flute and plays some very pretty restful tunes or perhaps sings some merry songs—and so the time glides by." And Amila "just enjoys everything with a lively keen interest, always cheerful, so that is a delight to be near her. She is a little, keen-eyed, sprightly body—her hair snowy white and her dear old face quite wrinkled—otherwise you could hardly think her to be 80."

The botanical couple was also happily planning their return to the Arizona Territory. Charles Crocker, the railroad magnate, had donated $20,000 to the California Academy of Sciences for research and discovery of Arizona. Better yet, wrote Sara,

he presented us with passes over his roads for the next six months. This gives us the freedom of several railroads up north, down South & across the continent to Ogden—also the freedom of the ferry boats from Oakland to San Francisco as often as we choose. This is a very great help to us. Mr. Crocker seems interested in our work as do many others, and in time we hope to get some good paying position where our bread will be assured.

We feel we are greatly favored and this on account of our interest in exploring.

Undaunted by the previous year's ordeals, they resolved to return to southern Arizona before the summer rains.

11

"Rushing, Reckless Life of a True Mining Town"

Southern Arizona, Summer 1882

→ DESPITE ALL THEIR PLANS, the Lemmons' intended Southwest exploration was delayed because Sara had illustrated so many specimens that she strained her eyes. The worst aspect of lying quietly in the dark for two weeks was that she had to deny her "chiefest pleasure": writing her Sunday letters to the family back home. In addition, JG felt weak and miserable all spring.

But by early May, they both felt well enough to travel, and the entire household was on its way to San Bernardino where Amila would stay with friends while her "wandering children" were exploring. Sara wrote Mattie, "We'll then go into Arizona, not in the region of the Apaches, but into the Huachuca (Wah-chu-ca) Mts where there is a military camp and the air is fine."

JG explained to their botanical colleague George Engelmann on May 3 that they'd be exploring just south of Tombstone because "the Apaches have possession of our old haunts in the Chiricahuas."

John Coulter, editor of the *Botanical Gazette*, reported in the June 1882 issue that "Mr. J.G. Lemmon and wife are off again for Arizona. One would have supposed that the experience of their last trip would have sufficed for a lifetime; but as long as a plant remains to be discovered, these intrepid explorers will try to find it."

Not anticipating how powerful the pull of the Huachucas would be, Sara and JG assured family and colleagues they'd be home in three weeks.

Getting to Tombstone proved harder than expected. Instead of pausing in San Bernardino a day or two, they stayed for two weeks while JG battled a recurring nervous fever and headaches. Finally they were back

Fig. 21. Photo by JG Lemmon of Sara preparing to draw *Hesperocallis* near Fort Mojave. The caption reads, "Desert-Lily: *Hesperocallis undulata* w/ Amabilis behind it to show relative sizes." Photo by author. Original at the UC and Jepson Herbaria Archives, University of California, Berkeley.

on the train, passing near Fort Mojave (now Bullhead City, Arizona) across the Colorado River from Laughlin, Nevada. Despite JG's weakened condition, they did pause to botanize. To their great joy, enough winter rain had fallen in the desert that the rare and stately ajo lily was flowering for the first time in eleven years. JG was thrilled to photograph it—with Sara behind the spectacular four-foot plant for scale. With its handsome white trumpetlike blooms, it was originally thought to be a kind of lily, but botanists have now reassigned it to the asparagus family. Its common name is not because it grows near Ajo, Arizona (although it does), but because the root bulb tastes like garlic—or *ajo* in Spanish. Ironically, possibly because of early enthusiastic collectors such as the Lemmons, the plant is now designated as "Salvage Restricted" in the state of Arizona, meaning any collection requires a permit.

On May 18 the couple reached Tucson, as Sara wrote in a hasty penciled note to the family: "Here we are again at a place that is fast becoming familiar from frequent visits. It is destined to be a large city. Since our last visit [a year earlier] it has grown rapidly, quite large brick buildings are going up, a new brick courthouse, city lighted with gas, etc."

The arrival of the Southern Pacific Railroad two years earlier had given the town a much-needed economic boost, and by 1883 Tucson would have three hundred establishments—including eleven saloons on Congress Street in the one block between Main and Church Streets.

During their previous stay in Tucson, Sara had discovered a cousin, Paul Plummer, who ran one of the few establishments that wasn't a saloon: the jewelry and optician shop on the San Agustin Church Plaza. The crossing of the Colorado Desert had been blisteringly hot at 108 degrees, and the Lemmons were grateful to visit and rest at Paul's shop. "Here it is 97 degrees, but fine and lovely," wrote Sara.

The next day they clambered aboard the train again and headed to Tombstone, which Sara described as "the best type of rushing reckless life of a true mining city on the outskirts of civilization": "In the mountains hard by are the Savages & wild beasts, here the civilized ones congregate, and *predominating* may be seen the finest types of our kind. Consequently, with wholesome restraint it is the best governed collection of inhabitants to be found, less real crime. On our way from Contention, the nearest R.R. Station, we took the stage—'Sandy Bob.' It was parked outside and in it—I the only woman—we met another return stage, almost all the occupants heavily armed—as they go out with [mining] treasure."

Because they still had another thirty-mile stagecoach ride to Fort Huachuca, the Lemmons paused for two days in Tombstone, staying at the Grand Hotel. It was a lavish establishment with fifty rooms, a wide inviting staircase with a walnut balustrade, elegant chandeliers in the dining room, a brand-new pool table, and a kitchen that included a twelve-foot range, ready to serve five hundred hungry diners.

Knowing how Mattie appreciated details, Sara described the furniture that had been shipped all the way from Michigan: "The parlor upholstry [sic] is scarlet-plush. Our room is furnished with neat white ash & silvered

Fig. 22. View of the San Agustin Church and Plaza in Tucson around 1880, looking up Camp Street from the Palace Hotel. Public-domain photo from National Register of Historic Places Continuation Sheet.

drophandles to the combination bureau & wash stand. Venetian shutters. Price $1 per day, Meals on European plan."

That evening she chatted with the proprietress, a Mrs. McBride, who'd owned the hotel only a few weeks because her husband, Archie, died of consumption the previous month. The ever-curious Sara wanted to see more of life in this true mining town, so she and JG took a stroll after supper, immersing themselves:

The air is ringing with distant and nearest sounds of music— instrumental and vocal, to lure the willing captives into the haunts of gambling & vice. There are no underground dens, these are reserved for taking out the treasure in the surrounding hills & mountains, but the dance girls sit side by side with the Negro minstrels, painted & decked in attractive colorful clothes, singing & chatting by turns to the most befuddled victim, whose pockets may have been discovered to have something therein worth the game, then anon encouraging another to "Put up some more, Jim," "That's the way to do it," with crisp terse oaths interspersed—I saw & heard these things through the cracks of the half open doors, during the space of about 2 minutes.

We hurried away for no one might surmise when a stray ball from some desperado might come our way. Almost every alternate door is an entrance to a saloon or gambling hall. The streets are thronged with men of all conditions, but few women are seen there.

Seeing the trainloads of "treasure" whetted her curiosity to visit an actual mine. The following day she continued:

We have been to Contention, "Tough nut," Girard, & Grand Central mines, all just outside Tombstone.

At the Grand Central we went down 600 feet, then threaded our way with torches into different cross cuts and through from what is called the old to the new works 1400 feet. The cuts are about 7 feet high and 6 feet wide, wooden tracks are laid, connect with a thin band of [illegible word], the small iron cars laden with ore to run upon to the cages & elevators. Terse signals are given out, and the engineer above who runs the big hoisting engine puts on steam and up comes the loads of ore, the same with passengers.

What a strange sensation as we went swiftly down into the pitchy darkness!

I looked up and could see a square of light where we started that looked about as large as a 5 × 7 pane of glass. This mine clears $50,000 per month. Two miners seem to be working as a rule together with each crosscut . . .

It was so stiflingly warm that the miners wear no clothing above the waist and the perspiration runs in streams down their faces and shoulders. They get for daily wages $4 and work Sunday as well as weekdays as the works cannot well be shut off as it takes so long a time to start up and is so expensive.

Their reward comes evenings at the gambling and dancing dens.

Ever frugal, she added, "The overseer says that as a rule they lay up nothing."

As fascinating as Tombstone was, the Lemmons hadn't come all this way to be tourists, and the next day they hired a wagon to haul all their camping and botanic gear to the Huachucas. The dry mountain air seemed to agree with JG, and Sara was hopeful that he'd continue to feel better—if he could be convinced to not overexert himself.

By June 2 they'd settled in at Fort Huachuca on the outskirts of what's now Sierra Vista—and learned they'd left Tombstone just in time: "Since then the whole town has been burned including our very good & newly finished & furnished hotel, 'The Grand.' Mrs. McBride, the widow of less than a month & proprietress of the hotel, rushing into the street, losing everything except the clothing that she had on, the insurance on the furniture being only one half—$6,000. She is left with quite a loss."

Fort Huachuca had originally been established as a military camp, but with the expansion of mining, the population in southern Arizona, western New Mexico, and the neighboring Mexican state of Sonora exploded. "Towns have sprung up as if by magic," wrote Major James Biddle, acting assistant inspector general of the Department of Arizona. All these new residents needed military protection from the Apaches, and he recommended that Camp Huachuca be expanded to a permanent post. Part of the rationale was economic: Housing troops long term in tents isn't as cheap as it might sound—canvas is expensive, it rots in desert sun, and in the long run, buildings are less costly.

Thus Fort Huachuca gained permanent post status in February 1882—just five months before Sara and JG arrived. Somehow in all the transition and construction, the couple's camping equipment was delayed, but once again the post commander, Captain Daniel Madden, who'd moved from Fort Bowie to Fort Huachuca, gave his own quarters to the botanists. Sara described their temporary home:

Four companies are already here & improvements are going on for the accommodation of ten Companies, 4 large barracks with stone foundations & two stories high for the homes of about 1000 men. New officers quarters are also to be built & all this accounts for the confusion and great lack of room.

So we feel ourselves quite fortunate in having two tents pitched for us, each 8 × 8 ft & a large fly between, as much canvas as is allowed for the Capt. of a Company. Then we have a cookstove furnished to be set on one side of the large oak tree under which our tents are pitched. This will help out in the culinary Dept. of the Botanists retreat. We shall be able to get up some little variety of food.

She also described the surrounding area:

The Government has assigned quite a large tract of land for this fort: 6 × 12 miles. This takes in some fine cañons, including the one that it occupies, all on the N.E. slope of the Huachuca Mts. Every cañon has a fine stream of water flowing from the Mts. down to the valley or plain below. I say below for we are a half mile up this beautiful cañon & at an elevation above sea level of about 6000 ft.

It is warm through the middle of the day—about as with you in June, but the evenings and early morning the perfection of temperature—no dew falls. Beautiful golden clouds form around the setting sun & sometimes the eastern sky shows clouds that look heavy with threatening rain & by the last of July the rains commence.

Even though they'd only just arrived and had missed the spring flowering season, they quickly realized they'd landed in an area of botanical riches—including a dozen species of ferns! They immediately resolved to spend the summer in the Huachucas to be ready for whatever plants would emerge after the summer rains.

Life near the fort wasn't as uneventful as the Lemmons expected—or hoped. Soon after their arrival, a band of seventy-five Apaches left their Mexican hideout and began moving toward the Huachucas:

In the first part of the last week we started out & spent the day exploring & a strange & startling surrounding was ours, which, had we known at the time, would have terrified stouter hearts than ours. We were not more than three miles & perhaps not one mile from a band of Apache Indians as we afterwards learned. A telegram was sent to the Commander that the Apaches were out again. Within four hours his troops were all in pursuit, with ten days rations. Three companies of cavalry & one of infantry left, only enough remained to guard the Fort. They were gone four days & could find no big Indian. The Commander keeps pickets constantly out twenty-five or thirty miles away.

Sara and JG wanted to set up camp four miles above the fort, but Captain Madden nixed that plan, advising they go no farther "than shouting distance in these troublesome times." Madden was all too aware of the danger: His eighteen-year-old son, a college student back East, had been on his way to visit the previous summer and was ambushed and murdered by the Apaches within 130 miles of Fort Bowie.

The Lemmons were amenable to the limit he'd imposed: "It seems almost lonely to be within shouting distance and overlooking the camp, especially on moonlight nights, just as it is rising, throwing half of our creation into shadow, converting tree stumps into Indians & all sorts of wild imaginings."

Sara derived much pleasure from the "culinary Department of the Botanists" and was especially pleased with the Diamond Rock No. 8 cookstove they'd been given. "I wish we had as fine an iron cooking range at home," she told Mattie.

> We have the exclusive use of it, upon which we cook cracked wheat, rice, hominy, beans, hot-cakes, apple sauce, tea, coffee & boma tea. Our bread we buy at the Company bakery. Seven 44 oz loaves for one dollar, just the same as is furnished to the officers & Soldiers & at same price through the kindness of the Commander, Capt. Madden. We can get any canned fruit & food and other goods at Gov. prices with 10% added as freight, just as the military through orders from the Commander. Then we are protected by the military, and if ill beyond our own resources, there is a post surgeon.

As the trip went on, she even mastered doughnuts and fried applesauce pies, to her husband's delight.

She also described their bedroom arrangements:

> In the matter of sleeping accommodations, we have two beds just stuffed with fine hay, paper is laid upon the ground, which is dry as powder, then these beds above that. We brought two pairs of blankets, and Capt. Madden has supplied us with two more so we are warm enough nights. As soon as they reach here, we are to have 2 single iron bedsteads. They will take us from the ground when

the rainy season comes on as we expect to remain here till after that time.

I cannot see how we could be better situated than we are for getting strong and well and in the prosecution of our beloved science.

Even at the very beginning of their Huachuca explorations, their "beloved science" was generating new and wonderful finds. On June 10, Sara sent Mattie a dried wild bergamot, *Monarda fistulosa*, a member of the mint family. When brewed, the tea tastes somewhat similar to Earl Grey. On the banks of a nearby stream was a showy columbine, *Aquilegia chrysantha*, a sharp contrast against a vivid backdrop of scarlet bouvardia and a deep red penstemon—both magnets for the dozen species of hummingbirds that frequent this mountain range. The botanists' camp lay in deep shade cast by three species of oak, "the lovely *madroño* with its red bark," and towering sycamore trees. Not far above them were three different kinds of pine trees. One, the Chihuahua pine, then called *Pinus chihuahuana*, she explained was pronounced "*She-wa-wa-a-na.*"

In addition to being chief cook, bottle washer, and associate botanist, Sara was the expedition's scientific illustrator. Almost all her work prior to 1882 has been lost, but most of her Huachuca paintings survived.

Sara and JG were in botanical heaven in spite of the potential danger:

One day last week Lemmonia and I started at 7 1/2 a.m. for a trip up our ravine and did not return till 6 p.m.

It was a splendid day—we took our lunch & flower press, umbrella & pick, also our sheath knives. When we had gone up the cañon about 3 miles, we suddenly met three of our Fort scouts— Apaches—all armed. They also had sheath knives. How repulsive they are to be sure! They wear their hair long, faces painted in red spots, sometimes interspersed with blue, legs bare, feet clothed with buckskin moccasins. They are very nimble, and run with long bounding steps. These three were cutting long poles to make bows of. (Perhaps we may be able to find some for the boys here. Will if possible.) They followed us some distance till I grew very

Fig. 23. Yellow columbine, *Aquilegia chrysantha*. Sara painted this watercolor in the Huachuca Mountains on July 4, 1882. Eventually it ended up with her grand-nephew, Harold St. John, who stored her works at his home in Hawaii. Amy St. John, Sara's great-great-grandniece, then donated the paintings to the University of California and Jepson Herbaria Archives. Photo by author. Original at the uc and Jepson Herbaria Archives, University of California, Berkeley.

uneasy, so we turned out of the trail and L. ordered them to go by a *vamoose*. So they went ahead & soon disappeared in the thick brush.

I felt uneasy enough for several hours, for although they are scouts and avowed harmless, I cannot trust to turn my back upon them. Many of those who are now hostile have been employed as scouts by Gov. and they are very good marksmen.

How easy it would be to fire as if by accident, when they are certain of not being observed!

At any rate we shall be on the lookout for them.

Not every hour at the fort was filled with work. Several officers were accompanied by their families, and Sara reported to Mattie, "Our social life is choice and most pleasant whenever we have the time." JG had brought the flute he'd hung onto throughout his Andersonville ordeal, and sometimes he'd play at sunset, the notes shimmering against the still air of the canyon. Often the music made Sara wistful, and she wished Micajah could join them.

The following week Mattie replied by describing her teaching duties, which made Sara even more homesick and sad that her family still hadn't met her no-longer-new husband. If only their railroad passes extended across the continent:

It would not be many months before you would see us with you. Your new relation would only be as glad as I to meet with you all. He now says that he is proud of his new sister's intellectual powers after reading the School Reports, & as for Father, he is sure he shall be glad to know him. I have boasted of his fine bass voice etc.

We write in love to George, the children, Father, & yourself.

Ever your loving sister, S.A.P. Lemmon

12

"A Botanical Paradise"

Southern Arizona, Summer–Fall 1882

↬ SHORTLY AFTER DAWN ON Sunday, July 17, gunfire rang
· out near central Arizona's Mogollon Rim and what's now the town of
Payson. Two weeks earlier, a band of Apaches led by the Tonto Apache
leader Na-ti-o-tish had fled the reservation. Angered by their losses at the
Battle of Cibecue Creek and resentful of reservation life, they went on
the warpath, successfully evading the pursuing troops while terrorizing
and killing settlers throughout the Tonto Basin.

Now, where East Clear Creek carves a deep canyon into the Mogollon's
cliffs, an Apache scout warned Captain Adna Chaffee of Troop D, Sixth
Cavalry, out of Fort McDowell, about an ambush set up by Na-ti-o-
tish. Chaffee had brought in extra soldiers and assigned them positions
that would block the Indians' escape route. Combat raged all day as the
seventy Apaches fought for their lives, and the screams of the wounded
and dying echoed against the rocks.

Finally at dusk, amid hail and thunder from a summer storm, the
Apaches retreated, having lost nearly two dozen warriors, including
Na-ti-o-tish. One U.S. private was mortally wounded and died at the
site. The Battle of Big Dry Wash was the last actual engagement between
the Apaches and the U.S. Army, but the war itself would continue until
Geronimo surrendered in 1886.

The next morning, unaware of the fighting 250 miles to the north,
Sara settled into the shade of the tent for her usual weekly Sunday "chat"
with her family—even though it was Monday. She and JG had been so
busy cataloging and labeling their plants she'd lost track of the days. He'd
hired a wagon and spent six days in the surrounding mountains, bringing

back heaps of specimens, some very rare. Next they planned to move their headquarters to Tanner's Cañon (now called Garden Canyon, a notable birdwatching site), ten miles south of the fort. They'd heard it was a well-watered area likely to be rich in undiscovered plants. Better yet, it was close to the Mexican line, and they hoped to do a short trip into the state of Sonora. She warned Mattie she'd be so busy collecting and sketching the perishable plants in watercolor, besides performing her usual camp duties, she'd have no time to write letters.

They hired a man to take them and their "bed, bedding, provisions & botanic fixings in a big wagon" to a deserted cabin far up the canyon.

Sara and JG weren't the only passengers on that big wagon. Several weeks earlier they'd botanized near the San Rafael Valley at the ranch belonging to the brother of Louis C. Hughes, editor of the *Arizona Daily Star* and future governor. While splashing through a swampy grove of willows, they scared off a mother turkey who left her five chicks behind. Three remained at the Hughes ranch, and Sara adopted the other two—despite JG's teasing comment, "We shall have to announce their place of burial in about a week."

I thought almost that way too, on the next morning as we took the little things from the box. They had lived in a fright all the day before, had no hovering, and would not eat and their wings dragged on the ground, eyes closed, gaping, with just a little weak piping sound.

I would not give them up but set about restoring them, held them close to my warm neck, kept my hands lightly around them, then took some cracked wheat, moistened & white bread with a sprinkling of ginger, opened their mouths and put the pellets down.

They revived a little.

Then L. caught two young crickets.

Voila! Two problems solved—not only did the chicks gobble every cricket offered, but the supply was unending. For days clouds of crickets had swarmed the camp, diving into dinner plates and burrowing into the couple's bedclothes. The chicks soon figured out they were on to a good thing: "This P.M. I find them so tame and hungry that I took them some distance from the camp and set them to work hunting their supper & crops full of grasshoppers, bugs and crickets. They follow me everywhere and knew my voice at the end of three days."

Sara named one Crick for the birds' favorite food and the other Vic because they were found on June 24, close to Queen Victoria's birthday. "So," she concluded, "They are brought still farther within the pale of civilization by having Christian names. How Lemmonia laughs at me, but I mean to win, & then I can laugh too!"

Now with a driver, a hired worker, and the wagon fully loaded with provisions and equipment—not to mention two well-fed turkey chicks— they headed up the canyon. Shadows deepened as the road narrowed between tall rock walls and crossed and recrossed the creek before plunging downhill to one ford so rough "the horses refused to draw us up the steep bank of the creek, & so in the middle of the swift-moving rocky mountain stream, more than the length of the body of the wagon in width, we were obliged to unload & pass to the other bank all our goods & then they still would not draw the empty cart.

"What to do?"

July is rainy season in southern Arizona, some years completely dry, others with mighty storms that pummel the desert and mountains. That year, 1882, was definitely wet. And—there but for the kindness of strangers:

A heavy shower threatening, lightning & heavy thunder startling us every few seconds—Just then a man came through the oak-shadowed cañon with two packed horses on his way to Charleston, ten miles on the plain.

He saw our fix and quietly said, "I think my animals can draw that wagon out."

In a twinkling our horses were taken out, his unpacked, the harnesses placed upon his animals & soon we were out—our goods hurriedly put back, his animals re-packed and the mysterious "Diamond Hitch" of the rope fastening everything securely.

With the storm fast approaching, Sara and JG could do no more than thank him fervently and offer him cold coffee and baked goods. The man, named only as Stewart, said he owned a ranch twelve miles away in Cave Cañon and went on his way, munching Sara's doughnuts.

The botanists had only one more mile and two creek crossings before reaching their cabin. But no sooner did they arrive at the next crossing when the horses balked again. By now the storm was no longer threatening: It had arrived, bringing torrents of rain. Again the exhausted, drenched humans unloaded all the gear, waded across the first creek with it, then the second, and lugged it to the cabin, which was another quarter mile away.

Once the wagon was empty, the horses were willing to pull it. Sara, normally softhearted toward all animals, commented, "It was a most trying situation, and if ever animals deserved severe treatment, it was then, but it would have been useless."

At last everything was unpacked, and once the two men were fed and headed back to the fort, Sara and JG rested on a stack of boxes. They concluded that all had ended well, even if the whole experience was trying at the time.

Two days later they realized how much worse it could have been:

After a heavy shower, we all at once heard a terrible rattling, unlike the thunder. Rushing out of our cabin & towards the stream, we saw a turgid mass of water sweeping down the stream, carrying trees, rocks & rubbish like feathers on its surface. In less than five minutes, a steady stream of this slush was rushing along in the creek, at least two or three feet deep, roaring and tearing along like a fury let loose.

Had we been overtaken by one of these sudden risings, we should have been borne along & drowned or crushed, & nothing could have saved us.

The flood brought to mind the story they'd heard the previous year of the two soldiers at Fort Rucker who'd drowned trying to cross White's River.

Pioneers, botanical and otherwise, often needed to rely on one another in the 1880s, and once again, Sara's people (and doughnut-making) skills didn't hurt:

> Just as we were looking in wonder at our volume of water, along came our friend Stewart on his return from Charleston, 10 miles across the plains, his two animals laden with stores. He had to betake himself & goods into what is called a "dugout"—here a cabin made by cutting into the side of hill and covering with a mud roof. This gave him protection for the night, and of course we invited him to take supper with us. We did all we could to show our appreciation of his former kindly services & parted company after the shower was over late in the evening, after all collecting around a big camp fire by the east end of our cabin to dry off & tell stories & so get into a comfortable state for the night.
>
> We were invited to his ranch twelve miles away in Cave Cañon and with the parting assurance that he would do all possible for us in our work.

Sara reassured the family, "We are both well. L joins me in love to you & all. We are having a fine season. No climate can be more delightful & salubrious than these mountain regions. The showers cool & sweeten the air & refresh the whole region. It is as pure & lovely as in June with you."

A week later, Sara wrote home using the stationery of the Tanner and Hayes Huachuca Saw Mill, one of three mills in the mountain range. Sawmills had been vital to the area because local wood provided both shaft timber and fuel for the silver mines. Once the Southern Pacific Railroad arrived in Arizona in 1880, so did competition from the cheaper California lumber mills. As a result, the Huachuca sawmill would close, as Sara described in a letter to her father:

As L. has gone two miles farther up the cañon on horseback &
left me here for a rest, my thoughts turn to you & home, so far
away. We are close to the Mexican line in an almost uninhabited
region as wild as picturesque. Mr. Tanner has a saw-mill here where
lumber is taken out and marketed to Harshaw and other mining
towns from 12–60 miles away. In about a week the mill is to be
moved, perhaps to Sonora, & then this place will indeed be the
scene of desertion & desolation with its 8–10 rough board cabins.
The military Reservation extends up here and so must be vacated.

After botanizing near the Tanner's Cañon cabin for a few days, the
Lemmons extended their explorations even farther up the canyon. At
dawn they locked the cabin they'd used as headquarters, left Crick and
Vic with enough food and water for two days, and packed up their one
horse with provisions and botanical collecting equipment. They were
thrilled to find not only the native potato but also Arizona's only species of
true lily: the lovely lemon lily, *Lilium parryi*, its scientific name honoring
their good friend Charles Parry. It's a spectacular tall plant—three feet
to five feet in height—and a deep buff yellow sometimes dusted with
maroon freckles. It's especially notable for its intense but delightful spicy
fragrance, which is not at all lemonlike, as the plant is named for its color
and not its smell.

Sara's comment was, "What a wild country this is to be sure!" Later
that day, she returned to camp on her own. The mill was shut and the
only sounds that of a solitary birdcall and distant thunder. The reason
for such a rich botanical season was the generous rainfall that year:

Every day or night there is a heavy thunder storm here & the rain
is copious. As I look at the flashes of lightning, much of it sharp
chain lightning, I think of your lessons that you used to give us
when little children—not to fear the lightning but calmly look and
admire the grandeur of the illuminated expanse. Here it does make
me shrink, sometimes.

I never remember to have seen anything so vivid in the East, or anywhere else as in Arizona.

By August 12, Sara and JG had moved back down Tanner's Cañon to the cabin, but they still couldn't tear themselves away from the Huachucas. JG was especially interested in that wild potato, which he hoped would prove to be a new species (Asa Gray later identified it as *Solanum fendleri*, which had already been introduced into cultivation by Charles Parry). JG had noticed that commercial potatoes were prone to rot, and while on the trip, he wrote to the *Gardener's Monthly and Horticulturist* urging readers to buy the tubers he and Sara would bring back to grow in their gardens.

In addition, starting in 1840 people of all social levels all over the world became obsessed with the beauty, variety, and novelty of ferns—there was even a word for it: "pteridomania," or "Fern-Fever." Collectors uprooted ferns so avidly that some species in Great Britain were nearly eliminated through their enthusiasm. Ferneries were established worldwide; the Dorrance H. Hamilton Fernery, part of the Morris Arboretum at the University of Pennsylvania, is the only freestanding Victorian fernery that remains in the United States. In 1829 Nathaniel Bagshaw Ward invented the Wardian Case, a mini-greenhouse to preserve and protect fragile ferns from London's coal-polluted air. The case eventually evolved into what we now know as a terrarium.

Not surprisingly, ferns had always been of special interest for both JG and Sara. The previous year they'd discovered a lovely bright-green fern with diamond-shaped fronds in moist pockets of granite rocks on the north side of a high peak of the Chiricahua Mountains near Apache Pass and Fort Bowie. JG named it for Sara, writing, "Dedicated to Mrs. Lemmon, whose maiden name is Sara A. Plummer and whose devotion to science, arduous labors and daring heroism while botanizing in the land of the Apache, entitles her to high honors and this timely recognition." It's still called *Woodsia plummerae*, or Plummer's cliff fern.

Earlier in 1882 JG also published a pamphlet called *Ferns of the Pacific Coast, Including Arizona*, which was well received at 35 cents ($3.50 for a dozen). At the same time, the Lemmons announced that Sara would be publishing an illustrated *Manual of the Ferns*. Sadly, that publication never happened, and all that remains is a description of the talk she gave on the topic.

So, given the international fern fervor, it's no wonder that Sara and JG kept an especially sharp eye out for those species in the Huachucas— and they weren't disappointed. On August 24, Sara described a hidden ravine JG had found. The thousand-foot walls were of syenite, then called "seinite," an igneous rock similar to granite, that provided a foothold for a treasure: "It was like a conservatory of ferns. There were 23 species, and three or four new to Arizona & new to us. They may prove to be new species—we hope so. This is one of our richest discoveries."

The two of them scrambled, slithered, climbed, and slid deeper into the ravine:

How still & dark and grand it is—shut in from all the world. It is probable that no human foot had ever trod its rock floor before, and for the first time it has yielded up its floral treasures. As Lemmonia pitched rocks down from the dizzy heights—how they reverberated. Nothing broke the gloomy stillness but the echoes of our voices and the sweet whistling notes of a little bird that kept flitting before & around us. At almost dark, we reluctantly left by the side entrance, discovered the previous day by Lemmonia, and hastened homeward, almost blinded by the vivid lightning, well paid for the effort.

One specimen they collected that day from "Conservatory Canon" still resides at the Chicago Field Museum. Only one of Sara's fern paintings has survived, and it's a combination of the cotton fern with coral bells.

Some days Sara chose to remain in camp, accompanied only by Crick and Vic, who'd matured by now into a pair of handsome young full-grown turkeys. They followed Sara everywhere, and "if they miss me, they will cry till they hear my voice. Whenever I sit down to write or paint at the cabin door, they come in with a hesitating mincing step and lie down in front, like two dogs."

Fig. 24. Sara's watercolor of *Henchera sanguinea* (coral bells) and *Notholaena ferruginea* (cotton fern), painted June 20, 1882, in the Huachuca Mountains, according to her notes on the back. Photo by author. Original at the UC and Jepson Herbaria Archives, University of California, Berkeley.

Fig. 25. Sara's watercolor of *Centaurea* sp., or spotted knapweed, painted August 14, 1882, in Tanner's Cañon. Photo by author. Original at the UC and Jepson Herbaria Archives, University of California, Berkeley.

Sara used those painting hours to practice the techniques she'd learned from Adelia Gates.

Yesterday I painted a pretty purple Gillia. This cañon is all aglow and alight with flowers. Especially at sunset is the display fine. The change from the sunlight to the golden bloom of thousands of a tall primrose is hardly less gorgeous, intermingled with the sunflower and a beautiful many-headed golden Grindelia [Arizona gumweed], the leaves of which turn edge-wise to the sun & glisten & shine from their resinous surfaces.

    Then, as if to vary the monotony of golden bloom, is the scarlet lobelia & sumach, toned down by rose-pink thistle and a large headed pink & white Centaurea, interspersed with wild grape, mixed with the oak & plane trees along the murmuring stream that supplies us with cold & delicious mountain water.

In a letter to Asa Gray, JG said they'd be giving twenty-five of Sara's paintings to the California Academy of Sciences. Tragically, most of Sara's artwork has been lost, quite possibly in the fire that gutted the academy after the 1906 San Francisco earthquake. But a few samples from their stay in the Huachucas have survived.

    In addition to fieldwork, cataloging the specimens, and painting many of them, Sara kept JG and herself fed—along with anyone else who showed up. While they were at the sawmill, a neighbor had shared a wild turkey he'd shot (not Crick or Vic!), and Sara concocted a turkey pie, complete with additional pork, gravy, and crust, all in a frying pan over the campfire. Side dishes and dessert included green corn with butter and graham cracker pancakes and doughnuts.

    Even though the abstemious Sara hastened to add, "We don't live that way all the time," she obviously relished the creativity required for camp cooking:

While at the *mill*, I cooked over 100 doughnuts for emergencies of botanizing, etc. In front of a good cookstove & plenty of eggs,

Fig. 26. Sara's watercolor (signed and dated on the back) of *Senecio Douglassii*, or Douglas' ragwort, painted June 1882 in Fort Huachuca. Photo by author. Original at the UC and Jepson Herbaria Archives, University of California, Berkeley.

Fig. 27. Sara's watercolor of *Jatropha* sp., most likely *J. macrorhiza*, often called limberbush or *sangre de drago* (blood of the dragon). The painting is signed on the front and labeled "S.A.P. Lemmon, Fort Huachuca Canon, Arizona, June–July 1882" on the back. Photo by author. Original at the UC and Jepson Herbaria Archives, University of California, Berkeley.

Fig. 28. Sara's signed and dated watercolor of *Ipomea thurberi*, or Thurber's morning glory, painted August 1882, in Tanner's Cañon. Photo by author. Original at the UC and Jepson Herbaria Archives, University of California, Berkeley.

Fig. 29. Sara's watercolor of a gentian, *Gentiana microcalyx*, painted September 13, 1882, in Tanner's Cañon. Photo by author. Original at the UC and Jepson Herbaria Archives, University of California, Berkeley.

Fig. 30. Sara's signed and dated watercolor of *Cnicus edulis*, or edible thistle (now *Cirsium edulis*), painted September 1882, in Tanner's Cañon. Photo by author. Original at the UC and Jepson Herbaria Archives, University of California, Berkeley.

Fig. 31. Plant identified by Sara as *Zinnia parviflora* in a watercolor she painted October 8, 1882, in Tanner's Cañon. It was more likely *Zinnia pauciflora*, now called *Zinnia peruviana*, or Peruvian zinnia. Photo by author. Original at the UC and Jepson Herbaria Archives, University of California, Berkeley.

I could not be improvident. Yesterday, as I was keeping the camp, I decided upon the novelty of baking Indian pudding & beans without an oven and only tin dishes—but as you know the Yankee blood will get stirred even in Arizona and so the ingenious brain set to work.

2 eggs, condensed milk, sugar, Indian meal [probably corn meal], salt & water. Made a starter for the pudding, parboiled beans, pickled pork, ditto for baked beans. Pudding was put in a tin can, that placed in a larger can with stones in bottom & water to boil up around it, covered close. Beans in tin can, covered close, & iron pail set over it. Slow fires of big sticks of oak. After eight hours, result: a pale face steamed Indian pudding—good. Beans ditto & to all intents with baked beans flavor but browned below instead of on top.

So when L returned we had a good day's showing for camp as well as for botany—besides I flaunted the purple Gillia before his eyes & on the whole while it thundered and lightninged on the outside, we lent a faint brilliancy from within.

And so life goes on here.

August 22 found the couple still in Tanner's Cañon, still lured by its botanical treasures but also contentedly trapped by the flooded creek's water "turgid with big boulders and vegetable debris."

Two days later Lemmonia returned from a rough day's exploration up a steep, rocky side canyon—injured. He'd scrabbled up into a "box" canyon where he could neither advance nor retreat safely. He chose to advance—slithering and bouncing down a precipitous rock slide— landing hard enough at the bottom that he broke the ring finger of his left hand.

Once again Sara's Bellevue experience proved useful, and "as there was no better means at hand, I played surgeon & bandaged it up in salt pork." The next day she splinted it and concluded, "The patient is doing well, only I fancy is inwardly writhing & figuratively weeping because he cannot use two hands, right in the midst of so much botanic work."

Despite the still-active Apaches, the Lemmons had decided their "beloved science" was worth the risk:

We are advised not to remain so far away from the Fort in these present Indian raids, but as the Apaches have never been known to visit down this canon, only to cross at the head of it beyond Tanner's mills, we take our chances. A prospector yesterday said that last week while he was over at his "prospects" with four visitors just over the line in Sonora, they saw 58 Apaches ride along the Santa Cruz Valley, not 400 yds from them, just before they killed 8–10 men, one woman & a child—Mexicans—at 4 P.M. They were well armed & on good horses. This was within 15 miles of us. . . .

I confess to a feeling of uneasiness, at times.

They never attack in the night. Seemingly they have some superstitions about it. Well, we hope to get out with our fine collection & on our way back to Oakland by the middle or last of October.

Sara and John loved their time in the Huachucas, and in some ways that summer and fall would be the high point of their lives.

Several weeks earlier they'd made friends with the four Duncan family brothers, and in September Sara and JG visited their ranch fifteen miles away on the western slope. A month later the youngest brother died of consumption, and one of the others rode the fifteen miles to the Lemmons' cabin, leading an extra horse so that Sara and JG could attend the funeral. The couple were so attracted to the Huachucas they even considered investing in the area, and the Duncans offered them both mining and ranching opportunities:

They hold out strong inducements for us to come here or to interest ourselves in mines & I have never seen Mr. Lemmon show so much interest in mining before. We hope to yet get a valuable interest in some of the mines in our travels about that will eventually lead to a comfortable revenue. . . . Then these good Scotch boys say that if we incline, we can put on stock with them as the ranch is large enough for thousands of head, & while we are exploring they will grow & increase. If we could start in with 50 head of yearling heifers, it would be a beginning.

Perhaps we cannot do anything. Time will tell.

On one visit to the ranch, these two middle-aged and supposedly frail botanists joined the Duncans in exploring a nearby cave:

Yesterday four of us on horseback, rode up their Cañon about three miles & at the foot of the high bluff of limestone near the water, fastened horses & with arms full of pitch pine, for torches, went up several hundred feet & down into a large labyrinthian cave, dark as a dungeon were it not for torches. There in this mysterious place we explored, climbing, crawling & walking for about two hours, coming across grottoes, stalactites & stalagmites and all sorts of curious lime formations. One compartment that we called a cathedral, with its altar, shrine & lofty fluted organ pipes, with a roof so high that we could not see the top with the help of all our torches & a big blaze on the altar. There were breakneck places & sometimes a crust so thin to stand upon that it would ring as we walked upon it like earthenware. Some of these foundation crusts would have large dark holes & down them we threw hard pieces of limestone that we could hear fall & tumble along within for yards.

What a change to the bright sunlight as we came out from this dark, weird place!

All too soon October frosts rimmed the water buckets each night, and the plants began to fade. The Lemmons' sojourn in this botanical paradise drew to a close.

13

"Considered by Less Ambitious a Fine Season's Work"

Oakland and Southern Arizona, 1882–83

→ NOTHING ABOUT FIELDWORK IN the 1880s was easy. In mid-October Sara and JG broke camp, packing all the camping gear, photography equipment, art supplies, reference library, and hundreds of specimens. Then came the challenge of finding transport from their Huachuca paradise to the Benson train station: JG rode several hours to locate a pair of mules in one place, a wagon in another, and a driver in a third location. After coordinating all the necessary pieces and loading up, they left soon after sunrise to catch the 4 p.m. train from Benson—only to discover that the railroad schedule had changed and the train had already passed through at noon.

Once again, local kindness came to their rescue. The station agent provided a canvas tent complete with wooden floor, bed, and wood stove. The San Pedro River winds its way near Benson, making fall evenings chilly, but the Lemmons settled cozily into their borrowed tent with a toasty stove, a full lunch basket, and a large package of mail from the Fort Bowie post office.

"We spent a charming unmolested interim," Sara reported to the family, "save one drawback—no bliss entirely unalloyed as sundry and many poets have sung . . . just as we were snugly ensconced for the night, were disturbed in the general stillness by a mouse running over our faces and nibbling in the direction of the lunch basket.

"A match, a light—it disposed of—another restful and satisfied silence."

But that wasn't the end of their nocturnal disturbances. When they heard another rustling, JG grabbed his slipper ready to hurl it at a bushy tail near the picnic basket. He paused, which was just as well as they suddenly detected the distinctive smell of—skunk. The couple wisely chose to remain silent and immobile:

> Such respectful and deferential stillness as pervaded our tent for the space of an hour till his highness chose to move himself out, leisurely nosing all about, sometimes brushing close by us.
>
> I leave you to fancy. Of course he was master of the situation, and we dared not give him a surprise or sudden start, every move must be open and above suspicion and few moves at that.

Having worried the night away, they slept in. After a 9 a.m. breakfast, they boarded the train and "left with rapid motion the scene of our season's hard labor, glad to be on the way."

Once back in Oakland, their days were filled with sending out plant specimens and cataloging.

In addition, their herbarium classes, particularly the ones on ferns, had taken off, and nearly every letter from Sara to Mattie included a request for more books and botanical texts. A new eight-volume text of Edward Joseph Lowe's *Ferns: British and Exotic* in 1883 cost $60 to $90 (now worth $1,750!), but Mattie was able to get secondhand copies. The books were exquisitely illustrated with wood-engraved blocks by Benjamin Fawcett.

Right before Christmas, as if life wasn't hectic enough, their landlord announced he wanted to use the building as a boardinghouse and they'd have to find another location for their entire household and herbarium.

The previous week JG had been flattened by "brain fever" so severe that Sara and Amila hired a carriage and took him to a friend's "luxurious home" to convalesce. The two women then masterminded the entire move to 1205 Franklin Street, a two-story house with six rooms, "a much better place," said Sara.

As difficult as their fieldwork had been, and as frenetically busy as Sara was, she still missed being out in the field. On January 2, 1883, JG wrote Asa Gray, "Mrs. Lemmon's health is always improved by breathing oxygen from plants in close conservatories or close canons, so she begs to go again to Arizona as soon as spring opens, but I argue that we *must* get off our present large lot of plants."

He pointed out she was still illustrating their plants and continued working with Adelia Gates, who referred to Sara as a "very promising pupil." He told Gray she'd painted twenty-five of the Huachuca species and that "they look very natural." Four days later, JG wrote Gray again and mentioned she was doing full-size illustrations and that they'd frame them before presenting them all to the California Academy of Sciences.

Throughout the winter and spring of 1883, Sara, JG, and Amila took turns being sick. Sara surmised their ill health was partly caused by the overwork of moving and maintaining the herbarium. She also speculated that after the Arizona warmth she and JG were struggling with Oakland's chilling sea fogs and cold breezes.

Of the three, JG became the most gravely ill. He wrote Asa Gray at the end of January: "My physician tells me often that I am not long for this world. Recently, of late my head aches terribly and my extremities become cold. It seems I cannot survive Andersonville: the cruelty of the rebels was not evanescent with the state of war. . . . I am so feeble from nervous prostration that I can hardly write, so please excuse incoherence."

Sara was so concerned for him that on June 1 she wrote Mattie she was afraid she'd soon be doing their work without her Lemmonia:

> I must turn more and more to the deep study of botany. The time
> may not be far distant when I shall have to move alone. I tremble
> as I think of the situation. Lemmonia is very frail this spring. Two
> physicians have talked with me about his case, and one expressed
> great surprise at his weakness. Said that he was quite shocked on
> taking a drive with him he seemed so weak & advises an immediate
> change of climate from the sea fogs to the dry mountain air. . . .
>
> I feel very anxious about Lemmonia. I dare not tell his mother
> as it might worry her too much at her age. So when you write do
> not say too much—in meantime I will keep you posted as to his
> improvement. I have great hopes for his getting better when he gets

up in the mountain regions as he always grows better. Of course each year it is harder for him to fight against the ills & so is it with all of us.

She'd also been worried by months of a nagging nighttime ache in her back and side. Physicians' diagnoses included rheumatism, or neuralgia, or maybe congestion of the liver—or perhaps the liver was actually ulcerated?

Finally she resolved to fix it herself. Sedlitz Powders were a commonly used laxative, consisting of tartaric acid, sodium bicarbonate, and potassium sodium tartrate:

> Two weeks ago I told Lemmonia that I had become tired of dosing without any relief, that I suspected that gall-stones might be the base of the trouble, that I was going to commence my own treatment. I placed a box of Sedlitz powders & a quart bottle of Olive oil at head of bed.
>
> After retiring I poured out a tumbler of oil (a coffee cup) and in about six deep full swallows, drained the glass, lay down quietly, no ill effects.
>
> In morning took a powder, breakfasted, and then the medicine operated. (If one powder won't do, take another.)

The result was two heaping tablespoons of gallstones, "one as large as your little finger to the first joint!"

In midsummer, despite their medical woes—and treatment—the little family sublet their rooms to recoup their twenty-six-dollar monthly rent. They then headed to Sierra Valley to visit Amila's other children: JG's two brothers and sister.

Sara reported that JG stood the journey well although he had to lie down twice a day. "Here we get not a sniff of the sea breezes in this high valley, surrounded on all sides by mountain peaks, wooded with pines & some of them distinctly marked with snow patches. The valley is a living green, grass a foot high. Not a rod from the house . . . is a fine mt stream from where we have delicious trout for breakfast every morning."

The couple stayed with JG's older brother, Judge William Lemmon, who Sara described as "unmarried, his hair quite gray. Very sedate, intelligent & kindhearted."

Judge Lemmon and Sara had corresponded frequently, in part because he was so grateful to Sara for taking care of his mother and brother. In fact, the previous month he'd offered all three of them a permanent home at his house.

Sara refused—politely. "I felt happy to hear him so express himself & feel grateful & all that, but Lemmonia & I have a work yet to do, we hope in our chosen and beloved science that we must not easily yield. If only a goodly measure of health comes to us—"

However, she did confide in William about a topic she hadn't discussed with her family. As frail and ridiculously busy as they both were, she and JG were determined to establish their own ranch—somewhere. Surprisingly, Judge Lemmon encouraged her, responding, "I was very glad to hear from you and that you are becoming settled on your own Ranch and believe it is one of the best of ventures. Anything in the shape of a Ranch upon which there is a living spring of good cold water will in time become very valuable."

He added wistfully, "I wish I could be with you one winter to aid in the planting—planting, planting, of more trees and vine, trees & vines."

Two weeks later, JG was still so unwell Sara wrote her father, "This is an anxious trial time for me." One helpful diversion was when the editor of the *Rural Press* sent them a seventy-dollar "instrument for the dry photography process." She promised to send sample pictures—but not before Christmas since they wouldn't have time to do the developing and printing before then.

While in Sierra Valley, she occupied herself sketching a few plants for various articles. This is also probably when she painted *Trifolium lemmonii*, the five-leafed clover that had launched JG's botanical career fifteen years earlier (see figure 12). During this trip she also painted an orchid, *Epipactis* sp., near Austin Creek in Sonoma County.

Fig. 32. Sara's watercolor of *Epipactis* sp., a genus of terrestrial orchid, is labeled on the front "Epipactis, July 1883, Austin Creek, Northern California." On the back in Sara's writing is the label "Not by Miss Gates but by Mrs. Lemmon." Austin Creek State Recreation Area is in Sonoma County near the town of Guerneville and next to Armstrong Redwoods State Natural Reserve. Photo by author. Original at the UC and Jepson Herbaria Archives, University of California, Berkeley.

Amazingly, by late summer both Lemmons had bounced back with sufficient strength to travel to the Arizona Territory again—at the most humid and uncomfortable time of year.

On August 26 Sara wrote Micajah from Cactus Ranch, six miles north of Tucson, where they spent three weeks exploring the Santa Catalinas. Some of that time they were accompanied by her cousin, Paul Plummer, the owner of the jewelry store in downtown Tucson:

> Mr. Lemmon, Mr. Paul Plummer and I started for a week's explo-
> ration in the most rugged of mountains. Went over a high and
> terrible pass through cactus, cat claws or a kind of thorny acacia &
> a kind of agave with thorns an inch long at the terminus of every
> leaf. These annoyances together with the numerous insects and rat-
> tlesnakes etc. made it one of the hardest trips we have ever made.
>
> We found some very interesting plants & so the compensation.
> The heat is quite terrific at this season of the year. Thermometer
> ranging high up in the 90s & it is about all we can do to get through
> with our everyday work & I find myself quite demoralized on the
> subject of letter writing.
>
> So you must pardon my meagre rambling scrap which I send
> more as an assurance to you of constant remembrance, no matter
> where we may be, than as a full report of our doings in full.

Despite the heat, rattlesnakes, and other "annoyances," Sara continued doing double duty as botanist and scientific illustrator. In addition, she was the expedition's medical expert: "Two days ago Mr. Lemmon got an ugly cut just below the instep from a sharp hatchet that fell from just above his head. It bled so profusely that I was frightened at first, but by binding it firmly in cotton & Vaseline, the blood was staunched & although it has laid him up for a few days, we think it all right."

JG also had an attack of terrible pain in his head—which she treated with mustard and Tabasco sauce.

Yet the benefits outweighed the hardships: "We were beside a fine continuous running stream of water, plenty of fish—chubs, delicious &

the first we had seen in Arizona. Deer plenty and fresh bear tracks close by our camp on the banks of the river."

Even far from home, Sara remained acutely aware of their shaky finances, which hung like a black cloud over their botanical adventures. In the same letter, she apologized for the delay in reimbursing Mattie, saying she'd misplaced the list of debts:

In the bustle & confusion I put it in one of my dress pockets, forgot to take it along, and then I could not send because I had not the money, but on our return in November will hope to have some classes & so earn it. We had so much illness last season that it took all we could raise to meet the extra expenses.

I would not state this, only as a just explanation for the delay & to say that I will then pay it with interest on or before January & do hope it will not inconvenience too much.

On September 9 the Lemmons were on their way to Fort Bowie where they once again benefited from a friendship forged two years earlier:

After spending three weeks in the Santa Catalina Mts, we came in to Tucson on Fri noon & started on the train for Ft Bowie at 7 a.m., next day. 109 miles by rail took us to Bowie Station, then by government ambulance we rode in & up the Apache Pass into Ft Bowie in company with Paymaster Major Clayton & nephew. The ambulance was sent especially for him, but we showed our credentials—pass & letter from Gen. Schoffield [*sic*: actually Major General John M. Schofield, General Crook's commanding officer] & he at once invited us to accompany them.

The Lemmons had been looking forward to reuniting with their friend Captain Rafferty who'd rescued them from their Rucker Canyon adventure, but he was away. So they stayed in his quarters. They'd had a rough time, Sara told Micajah and Mattie: "We have both been ill about 1/3 of the time, having been poisoned by drinking the bad well water at

Cactus Ranch in the Santa Catalinas Mountains. Lemmonia was so ill for two days & nights that I feared he would die. After that I was ill enough & it wore off in bad bowel troubles attacking us & even after we would think the trouble ended, again it would come on, following up till about a week ago."

They'd hoped to return to Fort Huachuca but were thwarted when they learned there'd been a case of yellow fever there—and that the patient had been moved to their beloved Tanner's Canōn.

Early Americans had good reason to fear any disease, but especially yellow fever. Nearly twenty-six thousand people had contracted the disease in New Orleans between 1839 and 1860. Because of the expanding networks of railroad and steamboats, the disease spread beyond the warm coastal cities to the nation's interior. In 1878, Memphis had fifteen thousand cases, and thirty-five hundred died. States reacted by creating local health boards that were basically quarantine agencies, and in 1879 Congress created the National Board of Health. The year before Sara and JG's 1881 sojourn in the Huachucas, a Cuban physician named Carlos Findlay had traced the cause to *Aedes aegypti*, a mosquito—but no one believed him. It wouldn't be until 1900 that his work would be acknowledged, and a vaccine was developed in 1930 by Max Theiler, who was awarded a Nobel Prize for his work.

Rather than risking the disease at Fort Huachuca, Sara and JG traveled to the northern end of the Huachuca range to botanize and recuperate at the Igo Ranch, owned by Margaret and V. (Vincent) H. Igo. Mr. Igo had been awarded the contract for grading the thirty-five miles of railroad bed between Benson and Nogales. One of Sara's surviving paintings was done at the ranch, and her penciled notes on the back of the painting include the comment, "We were guests of Brig. Gen Wilcox & wife & dghtr at Fort Bowie, Chiricahuas, and Huachucas."

On October 1 she wrote Micajah from the train station in Benson that they were on their way home, a month early:

We have had a rough hard season from droughts, fatiguing journeys & illness & return a month earlier than usual. Not using our

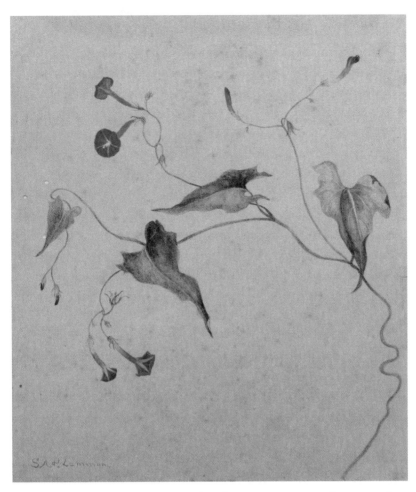

Fig. 33. Sara's signed and dated watercolor of *Ipomea coccinea*, or red morning glory or Mexican morning glory. She wrote on the back that she did the painting at the Igo Ranch September 9, 1883. Photo by author. Original at the UC and Jepson Herbaria Archives, University of California, Berkeley.

pass privilege toward the Gulf of Cal or down to Texas & New Orleans on account of our being so worked down that we are susceptible to any illness & the greater risk of catching yellow fever, going either way. . . . Many of our new & rare ferns alive & so bring back a good stock of living plants of about 13 species of the rarest, also seeds & pressed specimens of several of our newest discoveries, among them Tagetes Lemmoni, the shrub marigold, a shrub trumpet flower, a new species of onion, *Allium Plummerae* (to weep over) and our new *genus Plummera floribunda* that we only had a few specimens of last year but have a good supply of our immortals.

You see how your name is honored, the highest botanic honor that can be given to have a genus of plants. Mattie can explain. We come back, as the school girl says, *awfully* tired but very triumphant.

She added, "Mr. Lemmon has made several fine photographic pictures of trees, plants, & places & I have sketched in water colors some ½ doz or more of the plants, besides our botanic work which is by no means inconsiderable & would be considered by less ambitious ones a fine season's work."

By mid-October they were settled again back in Oakland, teaching classes and doing presentations.

JG described their year to George Davenport: "We have had a most interesting season of it, full of adventure, of peril, and of triumph. Now for another pleasant winter of studying up our trophies." Part of the "studying up" process included sending hundreds of specimens to Asa Gray.

For the most part, Sara accepted her role as the "& wife" partner in the Herbarium, rarely chafing against this description—until her New Year's greeting to Asa Gray on December 28, 1883. The Harvard experts had described one of the new species she and JG had discovered—without any attribution to her. She wrote that after musing if she should even bring the matter up, she decided she'd remained too silent for too long:

I was pained deeply, to be sure, so that for the time, my heart was too heavy for words, but the twinges are all over now—nothing chronic about the case. It was a touch of human weakness which is not foreign to poor human nature, is it? I frankly confess that the pain was inevitable.

I noted all the other collectors in the field, some of the gentlemen whom I knew had not explored as persistently or continuously, yet each record their proper credits, as recorded proofs of their labors.

It flashed upon me from the first that whether my name were Plummer or Lemmon, the work should be recognized like any other botanist or collector & when after three or more years of exposure to hardships, deprivations, dangers & all the concomitants that must needs come, the nonrecord pained me not a little. Our work is so united in the field, as well as out, that it seems like that of one—as indeed it is—till the summing up of the season's work.

In the busy engrossing life it had not once flitted across my mind that the plants so close to me in their wild homes would next appear in the books without me.

She went on to say, "It was a painful surprise, but let the matter drop."

She then discussed the detailed anatomy of creosote (*Larrea* sp.) and the couple's plans to continue exploring and finding new species for the next four or five years. After that, they would try to earn enough money to support themselves "for the rainy days ahead."

14

"Lives Cast in Pleasant Places"

Northern Arizona and New Mexico, 1884

→ EIGHTEEN EIGHTY-FOUR—THE SAME YEAR Mark
Twain's *Adventures of Huckleberry Finn* became a bestseller—began with
an optimistic note from Professor J. G. Lemmon in Meehan's *Gardener's
Monthly*:

> We are very busy this winter determining, packing, and sending
> out the large accumulation of years. Hope to finish by early spring.
> Thereafter will only keep authentic duplicates of all our collections,
> for reference. We expect to continue exploring Arizona and the
> border lands for several years yet, as long as health permits, but it is
> hard, wearing, dangerous work, and we may fail at times.
>
> I believe Mrs. Lemmon has twice the strength and determi-
> nation that I have. She has made dozens of excellent water-color
> paintings of flowers.

Little did they know how significant 1884 would be for them, stretching
their influence from the Pacific Slope all the way to New Orleans—and
beyond.

In spring, they received another year's worth of railroad passes from
Charles Crocker, still an enthusiastic supporter of their work. The prospect of free travel inspired them to form an ambitious plan: to botanize
from the Colorado River to the Rio Grande—from Needles, California, to
Albuquerque, New Mexico. They were also hugely relieved to have actual
credentials from the Department of Agriculture and the Department of
Forestry that would help open doors for them.

They set off in March taking advantage of the newly opened Atlantic and Pacific Railroad line that ran through Barstow, across the Colorado River, pausing at Peach Springs in the Arizona Territory, and then continuing on all the way to Albuquerque. Because their friends Dr. and Mrs. Ord had been transferred to Fort Mojave, just south of what is now Bullhead City, Sara and JG headed there first.

In 1858 Edward Beale had led an expedition—complete with camels—from Fort Smith, Arkansas, to Los Angeles. By then the Mojave Indians, who'd been in the Colorado River region for millennia, were fed up with the infestation of immigrants who'd been coming through by the thousands since the Gold Rush of 1849. Tempers flared, war broke out, and the U.S. government established a fort at Beale's Crossing to protect the white settlers. A peace agreement ended the yearlong conflict, and the fort is now a part of the Fort Mojave Indian Reservation. (The Pipa Aha Macav, or "People of the River," spell the tribal name with a "j" while the adjacent town is named Fort Mohave.)

Sara was impressed by the "gentle peaceable" Mojave Indians, writing to Mattie:

> The old women make ollas for holding water or water coolers, little miniature canteens, pitchers, canoes, etc. out of a fine red clay, paint them in their curious way & then bake them till they are tough & hard. These they sell for 25 & 50 cents. Others pound mesquite beans into meal for bread when they cannot buy or beg at the Fort, others plant corn & wheat along the low banks of the Colorado River by single kernels, dibbling little holes with a stick & often raise by the primitive means large & fine-looking fields of grain. They plant squash, melons, potatoes, beans, etc. and are not assisted by Government rations.

While based at the fort, Sara and JG spent ten days exploring the Calico mining district, northeast of Barstow, thanks to another friend who owned a mine there. Named for the mineral-rich and colorful mountain range, the area would become Southern California's most lucrative silver strike, eventually yielding $15 million. In 1883, the year before Sara and JG visited, the district produced 10 percent of the U.S. silver output.

Getting to the actual mines proved challenging for the Lemmons, not because of rugged terrain but because abundant winter rains had left the Mojave Desert "covered with many lovely flowers, among them for miles about was a beautiful desert lily. . . . It took about five hours to drive 10 miles, we stopped so often."

While in the Calico district, she told Mattie, they "visited several rich mines, threading our way through tunnels with candles in hand examining the 'hanging walls'—ore beds, etc. & botanizing above ground and becoming familiar with this curious mountain region."

Despite temperatures exceeding one hundred degrees, Sara was fascinated by the Calico lifestyle of the four hundred or so residents:

> Many of the miners live in the stone prospect tunnels where it
> is cool. Others have taken advantage of large boulders of spar
> & porphyry that are piled up in all sorts of forms like quaint
> architecture of towers, chapels, etc. with natural caverns. These
> they enlarge by blasting & make quite large rooms, very cool &
> comfortable. Some of them are cut out in such a way that they
> make a half dozen steps up & on a low shelf. They put a bedstead
> with spring mattress & so have *luxurious* apartments—below is
> room for a table, shelves, cook stove &c. This mining camp was as
> interesting to me as curious.

Ever practical, she was also intrigued by the fact that water had to be hauled from 2.5 miles away, and wood for the smelting was transported by mule seven miles.

They also visited the camp at Waterman's Mine, the first to be established in the district, back in 1880. One plant specimen, collected by Sara in the heat and dust near Waterman's Mine, was an endearing small purplish marigold, *Nicolletia occidentalis,* with the equally appealing common name of Mojave hole in the sand. She also gathered and pressed a monkeyflower, *Mimulus mohavensis,* carefully labeling it as collected in "Calico, Mohave desert." They sent the specimen to Kew Gardens, England, where it still resides in the botanical collection.

It was probably around this time that the Lemmons visited Mineral Park, another mining area just north of Kingman, Arizona. In among the plant specimens from this area that they sent to Harvard was a branch from an unusual shrub, a yellow-flowering snapdragon. Asa Gray identified it as *Penstemon antirrhinoides*, a species he'd first described in 1856. In 1906, twenty-two years after the Lemmons collected it, another botanist, Leroy Abrams, would disagree with Gray—and rename it *Penstemon plummerae*, in honor of Sara. Sadly, Plummer's snapdragon is yet another example of botanical nomenclature yielding to taxonomic accuracy: Scientists moved the plant to another genus, and it's now classified as *Keckiella antirrhinoides*, commonly known as the snapdragon penstemon or yellow bush penstemon.

The deities of botanical fortune continued to smile on the Lemmons as they moved east from Fort Mojave. On May 18 Sara wrote her father from "Peach Springs, Hualapi (Wal-a-pi) [*sic*] Reservation, West Central Arizona, 109 miles from the Colorado River on the west, 465 miles from Rio Grande river on East":

> It is almost impossible to remember where I last dated word to you, but it does not matter so long as you are assured that in the busy season of botanizing, you are always remembered.
>
> So far we seem to be unusually favored in every way & there never has been a season so favorable for collecting the flora. The rains have been very abundant & continue so. Even today there have been showers every hour or two accompanied by heavy thunder & I think at these times how you used to train us when children not to be afraid of the thunder & lightning.
>
> The remembrances of childhood keep me brave.

The Hualapai (the tribal name comes from *hwa:l*, meaning "People of the Tall Pines") had traditionally lived along the Colorado River near Peach Springs and Seligman in a ten-thousand-square-mile area of mountains, valleys, springs, and canyons. Once the Spanish and U.S. soldiers arrived, conflicts arose over who owned the land that had supported the Hualapai for millennia. Even though the Hualapai had served honorably in the early 1870s as government scouts against the Yavapai, an order was handed down from Prescott in early 1874 that all men, women, and children were to be captured. Soldiers then herded them into a corral in Beale Springs, what is now Kingman. After some weeks, in April 1874, they were told to start walking—south. They were then force-marched to a location on the river ironically named La Paz ("peace" in Spanish). After a year of unbearable heat, of beatings, and of starvation, those who remained strong enough escaped. Their journey took another year, but they arrived back in the Peach Springs area the following spring.

In 1875 A. P. K. Safford, Arizona's territorial governor, agreed that La Paz was an unhealthy place for mountain people, and he recommended that the tribe be given an area to call home. Eight years later in January 1883, President Chester Arthur signed an executive order establishing the Hualapai Reservation. It encompasses a million acres and follows a 108-mile stretch of the Grand Canyon's south rim. Unlike many reservations, there are no casinos; instead the tribe sells permits for big-game hunting, and the Hualapai River Runners are the only Indian-owned rafting company on the Colorado. The tribe also owns and manages the Grand Canyon West venue that includes the Skywalk, a glass-enclosed bridge with views looking four thousand feet below into the canyon.

Perhaps she was unaware of their heartbreaking history, but Sara disliked the Hualapai Indians, finding them very different from the Mojave tribe, and unfortunately disparaging them as "the dirtiest, sauciest & most degraded of natives." But she added their "bad" tendencies were "quickened by contact with the vilest of the vile of our own race that always congregate at important Stations along the frontier rail roads. There are seven buildings in a row at this little Station & five of them sell liquor over the counters. These Indians hang around, beg & steal & learn all the bad doings & copy as well as they can the vices of the dens. All the squaws old & young are badly corrupted by the white man, we are told."

Corrupted or not, the Hualapai women worked hard, she wrote:

The women may be seen any day by the half dozen, carrying on their back immense, heavy packages of green pulled grass that they gather, wending their way to the saloons & houses from the neighboring hills, dirty, black & ugly. They sell their bundles as they can get from 15, 25, 50 cents for teams & saddle horses & at once pass it over to the *bucks*, as the males are called. They seldom work, but take the hard-earned money & oftener spend it gambling or drinking in the saloons rather than in buying food for themselves & families. The men are most comfortably dressed. They constantly beg for old clothes & there being more white men than women on the borders, they stand the best chance.

This time the Lemmons were camped in a three-room former school-house a few yards from the railroad depot—a far cry from their usual tent nestled beneath a bower of peaceful pines. Hualapai youths had broken out the windows, described by Sara as "of good size made up with 4 panes of glass each 12 × 28 inches."

Hualapai trails crisscrossed both in front of and behind the cabin, and the Indians were no doubt bemused by the two middle-aged white people busily squashing plants at all hours. Face-to-face encounters between a Hualapai woman and the former New England-born Manhattanite didn't go well:

Some ½ dozen of them came down from the hills on the rear trail & were hanging about the back door. Lemmonia was down to the rail road tank for water, so I stepped out & told them to go away, then they came towards me & shouted "Bret, bret," meaning bread. I said "No, go away."

One saucier than the rest strutted towards me looking into my face in the most impudent way & said "Shoes, shoes." I never moved or winked, but fixed a silent stare into her face, then she made faces & grinned at me. Still I remained with a defiant stare.

Just then Lemmonia came up & they scattered, uttering the foulest dirtiest of English, but they have never been here since. We had previously learned that, if you give them anything, they grow saucier & more and more troublesome. Whenever we see one, young or old, coming near the door, we order them off.

Sara's antipathy toward these Indians, while probably typical of most white settlers at the time, may also have been partly because she confused the Hualapai with the Yavapai:

I must not forget to tell you that it was the Hualapi [*sic*] Indians who captured the Oatman family, killing the father, mother & some of the children, taking one of two daughters into captivity & they were pursued by the Mojaves who were friends to the white people. They bought the girls & returned them to the white people.

There was a book published several years ago giving a history of the affair.

Sara was most likely referring to *Captivity of the Oatman Girls: Being an Interesting Narrative of Life among the Apaches and Mohave Indians*, published in 1857 by Rev. Royal Stratton. According to the story, the Oatmans were part of a Mormon wagon train headed for California in 1850. The family had traveled on alone not far from what is now Yuma when they were attacked by Indians who clubbed the parents and four of the children to death, and kidnapped the two girls, Olive and Mary Ann. To many eastern Americans at the time, all Indians were Apaches: However, it was more likely the Yavapai tribe, not the Hualapai nor the Apaches, who captured the Oatman girls. The tribe traded Olive and Mary Ann to the Mojave Indians, who treated the girls relatively kindly. Eventually Mary Ann died of starvation during a famine, but Olive survived and, at nineteen, she was returned to white settlers at Fort Yuma. She went to college, funded by the sales of the book, married, and became a popular public speaker, dying of a heart attack at age sixty-five.

But the confusion about Indigenous tribes continued for decades.

In another letter several weeks later, Sara added, "Farther on we expect to visit the Moqui & Navahoe Indians. They are great Sheep raisers and make wonderful blankets that will hold water."

JG's March 1884 collection notes indicate that their Peach Springs plants were eventually distributed to the British Museum, the California Academy of Sciences, Kew Gardens, and Dr. Charles Parry. The couple also sent many dozens of specimens to Charles Edwin Bessey, yet another student of Asa Gray's, who'd left Harvard to become a botany professor at the University of Nebraska. The Lemmons' specimens from Peach Springs, as well as from all over Arizona and California, many with labels showing they were collected by Sara herself, still reside at the C. E. Bessey Herbarium at the University of Nebraska State Museum in Lincoln.

In late May their good fortune continued as they left Peach Springs and headed for Albuquerque. On the train they happened to meet a "wealthy, polished & appreciative" gentleman who invited them to use his ranch near Flagstaff as their botanical research center. His offer included "free use of horses, mules, plenty of food & good shelter." In addition,

He says that if we will make a complete set of the plants along the R.R. from the Colorado to the Rio Grande, he will see that we are well paid for them & it will be an object to us to do so.

This we make note of, you may be sure.

The gentleman was John Willard Young, a Mormon businessman and the son of Brigham Young, president of the Church of Jesus Christ of Latter-Day Saints. Five years earlier, in 1876, Brigham Young urged members of the church to move south from Salt Lake City and establish colonies in northern Arizona. John Young was among the two hundred who had heeded the call. He'd built a cabin at Leroux Spring, then secured a contract with the railroad to do grading and build ties, and constructed Fort Moroni as a timber camp in 1881. The stockade walls were to protect the workers from Apache raids, of which there were none. The fort was seven miles north of Flagstaff and is now known as Fort Valley.

The Lemmons would remain Young's guests off and on all summer as they traveled around northern Arizona and western New Mexico, botanizing in areas including Prescott, Ash Fork, and the San Francisco Mountains. Aided by their "trusty alpenstocks," they even made an astonishingly rugged ascent to the summit of Mount Agassiz. The view was so mesmerizing they lingered too long. Darkness fell soon after they left the top, and the two of them had to inch their way back down the rocks and ravines of the mountain, safely reaching the bottom hours later.

When not out in the field, the Lemmons were happily settled in a six-room cottage, situated in a large meadow amid a carpet of lush grasses interspersed with bright, strange flowers, as JG wrote later in a long and eloquent article for the *Overland Monthly*. "Everybody is kind," Sara reported June 29 in a postcard to the family and described the atmosphere as "delightful, the plain at an elevation of 7000 feet." That particular morning four antelope were feeding nearby.

On July 1, Sara wrote her father with a more detailed description of the "cottage," adding that it included a Steinway piano and "an efficient Texan woman as housekeeper"! "What do you think of that for the frontiers of a savage life?" she asked.

Among their forays from the cottage was a short trip to the Grand Canyon, described by JG as "a long broad, profound chasm 4,000 to 6,000 feet deep and over 200 miles long from east to west. In the abysmal depths of this chasm the always angry, turbid river, flecked below the frequent rapids with white foam, roars and falls toward its far egress." Surprisingly, they stayed only two days, probably because of illness, since their next stop was the nearby town of Williams, where JG had to rest from fever and bowel trouble.

That summer Sara would have been justified in feeling a small amount of unseemly pride, perhaps thanks to her letter to Dr. Gray the previous December. In July Gray published a huge book, *Synoptic Flora of North America*, and included the official description of *Plummera floribunda*, the daisy she'd discovered two years earlier near Fort Bowie and the comment, "whenever the name of Lemmon is cited for Arizonian plants, it, in fact, refers to this pair of most enthusiastic botanists."

Fig. 34. Sara's watercolor of *Castilleja integra*, or whole-leafed Indian paintbrush, which she labeled on the back as painted at "Kendricks Mt, N Arizona, 1884." Photo by author. Original at the UC and Jepson Herbaria Archives, University of California, Berkeley.

While botany was certainly her primary focus, Sara was willing to consider other possibilities in the Flagstaff area:

Now another scheme opens up. While Lemmonia botanizes for the next three months, I helping between times, can take charge of a little school here of a dozen or so of pupils and get $80 per month for the same, but this gentleman says that we can do better with our plants, as he suggests. Time will tell. At any rate we think it best to remain this season along the line of this R.R. as it is a new field & we cannot do better.

This is a rough country with all sorts of crude people & things but full of interest.

As it turned out, she declined the teaching offer. "Things but full of interest" included more Indian villages, including one that might be the Laguna Pueblo but isn't identified because the first page of Sara's letter has disappeared. But she described it as being two and three stories tall that required climbing up "by ladders that are of the rudest make & very old like the houses. When you reach the door-stoop & crawl in, for it is only about three feet high—after you get in, straighten up, for you are in a room six feet high. Little square 8-inch apertures to let light in, the ceilings & walls almost as white as though whitewashed."

Of course the head of the botanical expedition's "Culinary Department" would carefully observe how to cook traditional *metsene*:

Stones set up in this [fireplace] about the width of a hand & a large smooth flat slab laid on a fire built beneath, the stone heated hot. A batter, thin, of wheat ground fine [more likely corn meal], mixed with water laid on thin as a wafer, spread on about as one would spread paste in papering—this baked till the edges begin to curl up, then taken off & repeated perhaps a dozen times, the several sheets rolled & folded together & this forms *paper bread* or mat-sin'-ee. It is very delicate in texture and taste.

As an artist herself, Sara was also fascinated to see the Indigenous pottery. As early peoples became less nomadic, they began making utilitarian storage vessels for water, grains, or seeds. Gradually each society evolved their own designs and embellishment, and the pots became beautiful in addition to useful. Once the railroad arrived in the Southwest, it brought a market for Indian pottery, along with trading posts, fairs, and festivals.

Sara was intrigued with the pots that the women "decorated with all sorts of unique & curious figures, then bake it by heaping up cattle ordure, dry, all around it and making a hot smoke. This probably is rich in ammonia & is the only way they have of producing a fine pottery. I bought several articles & mean to take them to Oakland. I wish you could have some of them. They cost from 25 cents to $1.25 each, according to their size and work."

By June the Lemmons were in Albuquerque for the first of several visits. Sara was impressed by the town and proclaimed that even though it was only three years old with four thousand inhabitants, its many fine buildings compared favorably with those in Boston. The Atchison, Topeka, and Santa Fe Railway had arrived in town in 1880, and the first New Mexico Territorial Fair was held the following year—in a three-day downpour.

Sometime in September they happened to meet the president of the Territorial Fair, who urged them to display some of their plants as well as Sara's paintings at the upcoming October event. The Lemmons had been considering attending a much bigger event at the end of the year: the World's Industrial and Cotton Centennial Exposition in New Orleans, otherwise known as the World's Fair. Realizing that having displays at the Albuquerque fair would spread the word about their work and provide good advertising for the New Orleans World's Fair, they scurried back to John Young's ranch.

For the next month, they spent every spare moment from early morning to late at night preparing five hundred exhibits of plants, drawings, paintings, and even some of Sara's needlework. They then packed them all up for the train trip back to Albuquerque. There Sara set up the displays behind a guard rail and against a backdrop of "deep crimson cambric and windows curtained with white cloth to soften the afternoon sunlight."

Exhausting as it was, their efforts paid off. The last night of the fair, as tired as she was, Sara scrawled a postcard to Micajah: "We may say, in brief, that for us it was a perfect success." Her set of eight pieces of needlework took a blue ribbon, her watercolor sketches won first place with a "premium" of fifteen dollars, and their plant displays won many other "diplomas."

She concluded, "Our expenses were paid while there & it gave us a wider territorial reputation & also paved the way for the New Orleans Exhibition."

Inspired by their successes at the Territorial Fair, in mid-October they were back in Oakland and meeting with the officials in charge of planning California's role in the World's Fair. The discussions were successful, and on October 30, Sara explained to her father that she and JG would indeed be attending the fair, where they were "to Exhibit the forage, rare flowering plants, ferns & Field Sketches of the Pacific Coast."

Moreover, she wrote:

We shall hope it will give us a broader life in our work—perhaps meeting with many scientists whom we have long known by correspondence—but better still there will be plants, trees, etc. from all parts of the world. It is anticipated that the display in the Veg. Kingdom will be greater than the world has ever known before.

At any rate we shall have a large opportunity to study & learn much.

15

"Grandest Display the World Has Ever Known"

New Orleans and New England, 1884–85

→ THE NEXT TWO WEEKS were a sleepless flurry of sketch-
ing exhibit designs, lettering labels, cutting frames and backing sheets,
communicating with other exhibitors, carefully gluing down drawings,
finishing various paintings, and making countless lists of countless details.
The preparations for the New Orleans event were fiscal as well as logis-
tical: The state authorities had promised the Lemmons $600 toward
their expenses for the seven months they'd spend in the Crescent City,
plus free travel on the railroad for both them and all their displays. In
addition, the cost of all the carpentry to put the plants and other items
under glass would be covered.

The displays were no small enterprise. Sara and JG were in charge
of all the California exhibits, which included 250 species of flowering
plants, dozens of grasses, and seventy kinds of shells, along with examples
of silk culture and lacework. JG wrote their friend Sir Joseph Hooker,
the botanist and director of London's Kew Gardens, that they also had
one hundred bulbs of the stately ajo lily—and that Sara had made one
hundred other watercolor paintings for the fair.

The preparations were hugely stressful. Sara had completed the paintings while fighting a bad cold and a bilious attack. She was also frantic
about JG, adding, "I tremble every hour through fear that Mr. Lemmon
will break down under the pressure and anxiety."

But at last, on November 15, 1884, they boarded the Southern Pacific
Railroad train for New Orleans, leaving Amila to look after the herbarium and their living quarters. Sara reassured the family, "There is a nice
family who occupy two rooms upstairs . . . so Mother will not be alone or

Fig. 35. Sara's watercolor of *Platystemon californicus*, commonly called cream cups, signed by her on the back and dated 1884. This plant is a monotypic genus found only in California. Photo by author. Original at the UC and Jepson Herbaria Archives, University of California, Berkeley.

lonesome with friends within and a host outside." They'd also arranged to rent out two of their rooms to augment their regular rent.

Sara described the trip as "an interesting & most comfortable journey of 2495 miles by rail." Their four-day, five-night route wound its way from Oakland via Los Angeles, Tucson, El Paso, San Antonio, and Houston with a night of journeying through the Louisiana marshes. "We can hardly realize that we have been whirled over such an extensive country with so little discomfort & fatigue. The journey includes immense stretches of desert & plains with mountains often times barren or treeless, then again thickly clothed with oak & pine."

Of course this indefatigable couple wouldn't just sit back in their Pullman car, savoring the passing scenery. Instead they spent every day still carefully pasting plants on twelve-inch by sixteen-inch cards and wrapping up unfinished exhibits in time for the exposition's December 16 opening.

Upon arrival, once they'd supervised the unloading of their "number-less packages and quantity of Exposition baggage," including one thou-sand linear feet of plant displays and all Sara's paintings, they found their way to the fairgrounds. It was a madhouse, bursting with preparations for the big event, as Sara described to Mattie and Micajah November 30: "with its 1400 hammers clicking, workmen being directed hither & thither, carloads of articles being rolled, trucked, pushed & carried into the respective Exhibition places, the dense smoke of burning asphaltion [sic] for roofing, laying of iron pipes below, suspension of iron wire above, hundreds of busy people going in all directions—summed up presented a confusion that was almost dazing & bewildering, barring all possibility of description."

"The grounds look promising," she added, "and, although much is yet to be done in every direction, with a strong & efficient manager at the head of affairs, order will soon spring from the apparent chaos."

Once they'd seen all the exhibit materials safely delivered, the couple set off to house-hunt—on foot. After miles of walking, they at last found an affordable suite of rooms at the "most desirable" 422½ St. Charles

Avenue. The cost was twenty dollars per month, not including coal ("75 cents a *barrel*!" Sara fumed) for what would be their home for the six-month duration of the exposition.

What was it like being a New England Yankee and the loyal wife of a Union veteran in the Deep South? Sara described spitting—metaphorically only—on a monument of General Lee, commenting to Mattie that he "was no doubt a great general, but equally strong is the supposition borne out that he was a great & most responsible traitor to his country, so I have no respect for such."

New York's Democratic governor, Grover Cleveland, had been elected president earlier that month, narrowly beating Republican James Blaine—who'd carried Sara's adopted state of California. She also wrote, "We are now on intensely Southern ground & our election was solid south in result but most unfair in electoral counts. None of the Southerners so far look as though their votes were by any justice worth as much more than ours at the North."

But Sara had more immediate tasks than national politics to consider. Days before the exposition opened, the California commissioner appointed her as vice president of the Pacific Slope for the Women's Department for the Exposition. Her responsibility was to arrange for samples of women's work—inventions, paintings, lacework, and more—from California, Nevada, and Utah, all at incredibly short notice.

An oversight would eventually lead to what some believe is Sara Plummer Lemmon's most lasting legacy. By then most states had a representative flower or tree, but no one at the fair could agree on a state flower for any of the Pacific Slope member states. So, in "an unorganized inception," as a first step, the National Floral Emblem Society was born—with Sara at its head.

Then on January 1, 1885, she was appointed commissioner for California for the Department of Women.

"My Commission parchment is very handsome," she told Mattie on her official Department of Women's Work stationery. She then added, "All this gives me much to do. . . . As Vice President I must have a general oversight of all articles or goods from Alaska, Washington Territory,

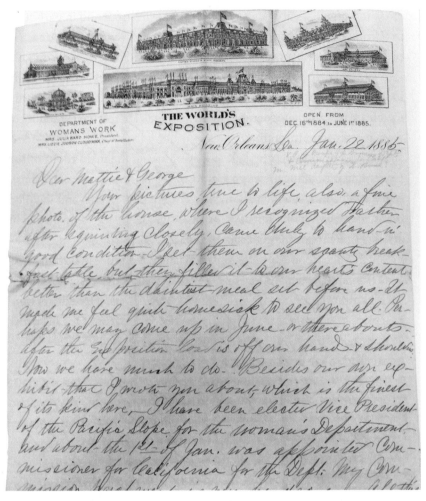

Fig. 36. Sara's January 22, 1885, letter to Mattie and George, explaining her new role at the New Orleans World's Fair. Photo by author. Original at the UC and Jepson Herbaria Archives, University of California, Berkeley.

Idaho, Nevada, Oregon, Utah, California, Arizona & New Mexico & letters to write in all these directions. . . . This is indeed an immense and wonderful Exhibit—Mattie, you and George at least *must* come down. Excursions are cheap, and such a chance to see the country and this *big* exhibition will not occur *here* again."

JG too urged his in-laws to come see the whole fair and in particular to admire the handiwork of his "dear little wife," saying, "Her space is really a Bower of Beauty with its sky blue ceiling, its gilt walls divided into large panels by glass-covered botanical specimens—each a study looking like exquisite tile-decoration around each of the squares, within which are placed chef-d'oeuvres of the painter's art and the evidences of woman's skill and invention."

Sara had used the entire balcony overlooking the main room of the Government Building to design a handsome gallery that contained waist-high glass cabinets, which showcased the smaller, fragile, and detailed items, along with taller displays for larger works. In all, she managed to arrange two thousand objects in the space.

She may have come into the position late in the process, but the *Oakland Tribune* reported proudly that the city's own "energetic and practical" Mrs. Lemmon performed work that was "marvelous." Her efforts were well reviewed locally as well: The *Times-Democrat* of New Orleans wrote March 29, 1885, that "in point of arrangement the Pacific coast work is unexcelled in the building."

The *Times-Democrat* writer wasn't shy about Sara's own work: "In fact, Mrs. Lemmon is represented in the Exposition by her work as a scientist, artist, botanist, lace-maker and needle woman. Her collection of water-color pictures in the department are worthy of close study, embracing almost the entire wildflower kinds of the Pacific coast."

The *Oakland Tribune*, again justifiably proud of its own, added on April 11, 1885: "One of the most delightful of Mrs. Lemmon's many talents is her artistic ability. In the art of delineating the graceful forms and delicate colors of plants and flowers, she stands without a peer. As Audubon painted with fidelity and spirit the myriad forms of animal life with which the earth is covered, so has Mrs Lemon [sic] transferred to paper the plants of the earth in all their native freshness."

Sara and JG spent a busy and dynamic seven months in New Orleans. In March she apologized to her father for not writing more often, explaining that goods were still coming in every day from California,

1982.127.212

Fig. 37. Sara's handiwork at the display of the Pacific Slope Women's Department at the New Orleans World's Fair. Photo courtesy of the Historic New Orleans Collection, Williams Research Center, New Orleans, www.hnoc.org.

ensuring that "the 1600-foot ground floor and walls, screens, showcases, etc." would be well filled for the duration of the fair.

The couple's days started with "nice coffee, graham bread (five cents a loaf), hash or fried oysters, fresh butter, fried apples." Each day Sara packed a lunch, consisting of bread-and-butter sandwiches with fruit, for the two of them before they strolled one block to ride a streetcar, for another five cents, to the exposition grounds. She had her own office with a little alcohol lamp to heat up tea to accompany their lunch. Her Yankee frugality prevented her from buying any of the expensive fairground food—but they made an exception for popcorn.

At night she often prepared a hot oyster stew, serving it with crackers, bread and butter, and "nice tea." After all, butter was only forty-five to fifty cents a pound, and oysters could be had for fifteen cents per dozen.

As jam-packed as her schedule was, they made time to go to classical concerts in the city, and she even organized small impromptu musicales in the Woman's Department to draw more visitors.

One of Sara's favorite fair exhibits was Maine's, with its plush lap robes, manufactured in Westbrook. Often she paused nearby, just to inhale the smell of the pine, birch, and other woods that brought back childhood memories with one sniff.

She also described visiting the exhibits of "Japan, China, Russia, British Honduras, Guatemala, S. A., etc. The display of sleighs, sleigh robes, furs of all kinds, malachite tables ($1500 each), all sorts of malachite nic-nacs [sic] is very interesting. I cannot begin, if time would allow, to tell you of the general appearance of 18,000 different exhibits. It is yet simply bewildering as a looker or, how much more so to attempt a description by such a poor delineator."

Sara and JG's time in New Orleans raised their social profile more than either would have imagined and led to connections that would benefit them for years. Back in her New York days, Sara's circle had revolved around artists and activists, and while at the fair, she undoubtedly renewed some of those contacts as well. She also worked closely with Julia Ward Howe, the activist, author, and poet who had written "The Battle Hymn of the Republic" some twenty years earlier.

But the people the Lemmons hobnobbed with most often were Clara Barton and her field agent, Dr. Julian Hubbell. Clara, who'd founded the Red Cross three years earlier at age fifty-nine, and Sara had much in common: They were both New England Unitarians, both had worked as teachers and cared for wounded Civil War soldiers in New York, and both had worked with the Sanitary Commission during the war. Clara and JG had much to talk about as well, since in 1865 she managed the Office of Missing Soldiers, tracking, identifying, and burying the remains of thirteen thousand soldiers from the Andersonville Prison.

One can imagine the lively chatter every time they got together, and the two couples made plans to visit one another on both coasts.

By May New Orleans was warming up, and Sara said, "The climate is becoming rather enervating to us high toned Californians, accustomed to the fine sea breezes." The closing of the exposition was near at hand—which was a relief, since both Sara and JG were exhausted and thin.

Packing up the exposition displays was an immense job that wasn't complete until mid-June. Although eager to see Sara's relatives, the couple then traveled north slowly, stopping to visit the Andersonville and Florence prisons. From there they headed to Louisville and, according to a postcard from Sara, "the 'blue grass' region where the wonderful horses are raised."

They then went by train to Washington, where they were able to relax with Clara Barton and Dr. Hubbell without the daily pressure of the fair responsibilities, before traveling on to Philadelphia and New York. By September they'd arrived in Cambridge, where they rented rooms on South Mitchell Street and were able to work with Asa Gray and Sereno Watson at the Harvard Herbarium. Sara had caught a bad cold but said they had such a splendid time anyway that it was hard to drag themselves away. She exclaimed in a postcard to her family how wonderful it was to compare notes about the plants they'd brought and to actually work in person with the herbarium staff after so many years of correspondence.

But at last they were on their way to Dover, and Sara finally—finally!—introduced her family to the man she'd married four years earlier.

Sara and JG had a delightful time with Mattie and Micajah, chatting, exploring the farm, touring local attractions, revisiting old haunts, and reconnecting with neighbors and friends. They all spent evenings together reading and reciting poetry, and singing "Yankee Doodle Dandy," accompanied by JG on his little silver flute.

The only shadow in the otherwise sunny family gathering was Osgood, Sara's oldest brother. On October 10 he wrote a cross and disapproving letter to Mattie about the "poor and plucky" Lemmons and how their botanical enterprise was "perfect nonsense." The real source of his irritation appears to have been money that he and others had lent to Sara and JG, contributions "that would be used to tramp among Indians, roughs, and dangers that are simply wicked to incur, just to follow a morbid idea to acquire reputations for finding new species of grass and plants. . . . I would recommend some of the pluck be used in winning bread, rather than tramping in wilds and mountains and going hungry for fame."

Sadly, Sara and JG's warm and loving visit in Dover ended abruptly and much too soon. A telegram, probably sent by Adelia Gates, arrived, stating that Amila Lemmon was dying. Sara and JG packed hurriedly and left immediately for the West Coast, hoping to make it back to Oakland while she was still alive.

16

"Our Hillock in Cholame"

Near San Luis Obispo, California, 1885–87

→ SARA AND JG WERE twenty miles west of Carlin, Nevada, when the conductor delivered another telegram to them. Amila Hudson Lemmon had died.

By the time they arrived home, she'd already been buried.

They consoled each other, saying that at least Amila had died peacefully with friends and family taking turns at her bedside. Even their good friend Ina Coolbrith, who would become California's first poet laureate, had sat with her—while writing a poem in Amila's honor.

"How we hourly miss the dear little Mother," wrote Sara to Mattie. "We shall never see her like again." She also mourned that their New England visit was cut so short: "Hope Father stays merry & well if he sings Yankee Doodle every day till we come again, which we hope to do in three to four years."

Gradually, the couple settled back into life at the Lemmon Herbarium without Amila's sprightly presence—starting with a deep cleaning. The Franklin Street house needed much attention: It hadn't been thoroughly scrubbed in three years, everything was coated in dust, and even the rivets had fallen out of the door. They (mostly Sara, despite more bilious attacks) washed the walls, scoured and painted the floors, repaired the broken furniture, and sent the carpets out for steam cleaning.

In spring JG was appointed California commissioner of forestry, which was accompanied by the first steady income they'd seen. The money was especially welcome, as the World's Fair finances were so poorly managed the exhibitors felt fortunate to receive half of what they'd been promised. Sara and JG were also still trying to pay Amila's medical expenses, since the doctor had visited her two to three times a day for seven weeks at $2.50 to $3 per visit.

Then came more bad news: Their landlord was selling the house to the railroad, and they would have to move—again. Well-accustomed to packing and not wanting to house-hunt during the field season, they put the entire Herbarium in storage and set out on their summer travels.

Their first stop was three hundred miles south. Years earlier in Santa Barbara, Sara had become good friends with the Edgar W. Steele family at their Corral de Piedras ranch in Arroyo Grande, near San Luis Obispo. Once they'd met JG, the Steeles had approved heartily of the Lemmon marriage and were appreciative and enthusiastic supporters of the couple's botanical work. The two families remained connected over the intervening decade, and San Luis Obispo became a frequent stop for the Lemmons. While staying with the Steeles once, they'd even tried to start a botany class in town but weren't able to attract enough students. They spent so much time there, however, that the Steeles converted a spare two-room cottage to a temporary herbarium. All the hosts asked in return was for Sara to tutor their young son Eddie in spelling, reading, and drawing for an hour and a half a day—an assignment the former teacher was happy to complete.

In July Sara and JG returned yet again to the now-annual Chautauqua Literary and Scientific Circle retreat at Pacific Grove, where Sara gave several talks on both botany and ethnology. Next they headed to Santa Cruz so that she could paint a mariposa lily.

Both Sara and JG were members of the Grand Army of the Republic (GAR), and Sara was active in the Lyon's Post Woman's Relief Corps. Each year the GAR held an August encampment, and in 1886, probably due to urgings by the Lemmons, the organization invited Clara Barton to be accompanied by her "well-beloved" field agent and constant companion Julian Hubbell. (Speculation no doubt swirled around Clara and Julian: She was twenty-six years his senior and eventually even deeded her house to him before she died. Yet there's no evidence that she regarded him as more than an especially valuable assistant.)

After the event the two couples traveled to Webber Lake, where they spent a merry and memorable two weeks together. No cabins were available, so they created their own camp, settling in under a circle of pine trees at an elevation of 7,860 feet. They ate heartily, roamed the nearby mountains, and celebrated Sara's fiftieth birthday. They took daily excursions to nearby sites, including Indian-pictured rocks and Mount Lola.

Every evening ended with JG playing "Yankee Doodle" on his little silver flute, and Clara always asking for an encore of the "Nicodemus" song—most likely "Wake Nicodemus," the 1864 song by the abolitionist Henry Clay Work. JG told Mattie in a September 3, 1886, letter:

> Best of all we have had stories and songs and recitals about the camp fire, tales of the Rebellion and of the European war that will ever remain in memory since they are told so well and excited such interest. . . .
>
> Miss Barton is certainly the best conversationalist I ever heard. Her stories are well selected and artistically told.

He added that Miss B and the "delightful little Doctor" were the most agreeable campers he and Sara had ever encountered, equal to any emergency or challenge, including a forty-mile horseback ride—eleven hours in the saddle! Clara loved the whole expedition as well, writing years later about their time at Webber Lake "to which my heart memory always clings."

It wasn't all idle play, however: Sara wrote of sketching the different species of evergreens so intensely she gave herself a stiff neck and headache.

Despite all the camping, hiking, and riding, Sara and JG's health was still a constant worry. JG's army doctor had diagnosed his consistent chest pain as heart trouble, and Sara still struggled with the "torpid liver" blamed for her bilious attacks and neuralgia. Still, she wrote Mattie, "Death will surely come to us all in some form & there is no use to feel in dread or die daily in worrying. I believe in keeping cheerful with a lively interest in the world about us and at large, and that always seems to keep one longer alive."

Fig. 38. Sara's watercolor of a "Great Sugar Pine" (*Pinus lambertiana*) as reproduced in the Lemmons' book *How to Tell the Trees and Forest Endowment of Pacific Slope*. Sara's signature, "Mrs JG Lemmon," can be seen tucked above the right-hand pine cone. Photo by author. Original at the UC and Jepson Herbaria Archives, University of California, Berkeley.

Fig. 39. Sara's watercolor painting of a narrow-cone or knob-cone pine (*Pinus tuberculate*, also called *P. attenuata*), reproduced in black and white in the Lemmons' book *How to Tell the Trees*. Photo by author. Original at the UC and Jepson Herbaria Archives, University of California, Berkeley.

Sara also still believed fervently in being fiscally prepared while remaining alive. In her continuing effort to set aside money for old age, she'd long been interested in land investment. In addition to the building she'd purchased for Santa Barbara's library, she'd bought two plots in Santa Barbara that she'd hold on to until 1909 and 1910. Years earlier JG had purchased a claim in Sierra Valley that he sold in 1884 for $500; that money helped defray their expenses in New Orleans, supported Amila—and even paid $80 toward her funeral costs.

Now, during their frequent stays with the Steeles, and like so many other newcomers to the San Luis Obispo area, Sara and John fell in love with the surrounding land. They spent days wandering the mountains and rolling oak-studded grasslands by wagon and camping in their tent at night.

Again and again they talked about their dream ranch and homesteading here—but how would they pay for it?

Sara tried her father first. On November 10, she wrote him that government land was available for $1.25 to $1.50 per acre for 160-acre parcels. This reduced price was made possible by the Homestead Act of 1862, signed by President Abraham Lincoln, as a way to get more land into the hands of American families.

Because everyone anticipated that the railroad would soon move into that part of California, the value might rise to $25, $30, or even $40 per acre within a few years. Needless to say, there was considerable interest in the snowy Northeast, and Sara warned her father that thirty thousand people were said to be on their way from the East Coast.

"Now you see, Father," she wrote, "I am speaking two words for myself—as I know it would never do you much good to get hold of that much, but it would me, it would give us a good help ahead and not cost you much trouble perhaps & I should be so glad to get hold of some such thing *if* it can come to you, and you would be willing to get it into my hands."

Micajah said no.

Mr. Steele, however, had repossessed a defaulted claim near Cholame, about sixty miles east. He offered to lend Sara and JG the money for it—at one dollar per acre—promising that no repayment was necessary until the Lemmons' "ship came in."

So, in late 1886, JG filed a claim on 160 rolling acres in Cholame, thirty miles from the then-famous mud baths of Paso Robles and overlooking the site where, sixty-eight years later, the actor James Dean would die in a car accident. Knee-high grasses, golden and "powder dry," carpeted the hills in fall, said Sara, but the rains would bring new grass and flowers "like magic." To their delight, while exploring their new plot, they discovered an accessible spring of sweet water—highly unusual, since most residents had to have potable water brought in by wagon.

Still the farmer's daughter, Sara described their land to Mattie: "Fully 100 acres are fine plowing land—fine for wheat or any grains. Other portions would be perfect for fruits such as olives, grapes, figs, apples, cherries, etc. In places where water has cut through the lands we cannot see anything but rich soil for several feet."

All that was required was to live on the land for six months, make some improvements, and then pay an additional $1.25 per acre, a process called "proving up." The land would then be theirs.

In November they settled in. They chose a spot near the spring in the center of the plot and, for fifty dollars, ordered enough lumber for a ten-foot by twelve-foot cabin. Yes, it would be tiny—but at last, instead of renting, this aging couple would own their very first house.

In the meantime, the couple continued camping in their canvas-walled tent. Winter temperatures in central California can dip near freezing, so they'd brought "four pairs of heavy blankets, comfortable feather pillows, night clothes, extra wraps for day & night, rubber coats, leggings, & overshoes, extra shoes, stockings," along with "cooking utensils, a large box of provisions, cooked & uncooked, apples, dried fruit, graham crackers, pickled pork, beans, &c, &c. Satchel filled with toilet articles, towels, medicines and numberless little comforts, quite a large supply of stationery, &c, &c."

They were ready for whatever winter would bring.

Until disaster struck.

On Christmas Day, 1886, Sara wrote Micajah about "a little ill fortune that befell us" two weeks earlier. They'd hired a man with two horses and a wagon to drive them to a wooded area 2.5 miles away to gather oak firewood. The first day Sara carried their valuables with them but then worried that she'd lose them. The next morning, she carefully hid them in the tent before leaving to go get more wood.

> While out the second day, our whole equipment was destroyed
> by fire. How, we know not. Someone passing on the trail may have
> carelessly dropped an end of a cigar stump or lighted match as
> our own little campfire was especially cared for. A spark would be
> enough in such tall dry grass. Suffice it, we found not a shred of
> tent or bedding or clothing. We lost over $60 in coins besides all
> our traveling passes.
>   This is heavy upon us and will cramp and inconvenience us.

The biggest heartbreak was finding the melted fragments of JG's little silver flute among the ashes. "We *did* lift up our hands in agony," Sara wrote.

Even worse, the fire "was creeping rapidly up the hills, threatening other places." They quickly moved the horses out of range and

> went to work fighting fire with some wet grain sacks. Several people
> joined us from eight miles away. At dark the fire was subdued, we
> having worked for over two hours. We had eaten nothing since
> morning, were fatigued, dispirited, dripping with perspiration, not
> an extra garment or blanket or bed or supper & nowhere near to go.
>   What to do?

They brought the horses and wagon back, unloaded all the wood they'd spent the day collecting, and hurried down the hill to the post office a mile away. The postmaster kindly lent them two quilts, and they drove an additional six miles to the Rockwell family, who happened to be related to the wife of Sara's cousin, Paul Plummer. Even though it was late, the family "welcomed us, gave us hot drinks, a good supper & put us to bed & O how soft & warm & comfortable the bed seemed to our aching bodies!"

The next day the Rockwells lent them five dollars for travel money and a couple of blankets so that the borrowed quilts could be sent back to the Cholame postmaster. Sara and JG then traveled on to San Luis Obispo and the ever-hospitable Steeles, where "I gave out & could not sit up much for two days, but am on my feet busy packing to start for Oakland."

Determined as ever, Sara and JG returned to the Bay Area, but just long enough to replace their camping gear and gather enough supplies to last the six months. Everyone they knew pitched in with financial and practical donations: The Lyon's Post Woman's Relief Corps Auxiliary presented them with three handmade "comforts," and "Mrs. Dr. Buck"—Dr. Annette Buckel of the Ebell Society—sent them off with a dozen cans of Eagle condensed milk.

Predictably, Osgood was outraged that Sara was irresponsible enough to "leave money and valuables in a tent for hours subject to the handling or theft of any loafer who might chance that way." Yet, even he sent a check for $100 (equivalent in 2020 to around $2,500). Sara thanked him effusively, saying the gift "bridged over the chasm of our ill luck with a good solid plank."

The couple returned to Cholame, and, despite few tools and even less construction experience, started to build. By March 1 a small sturdy cabin stood proudly on the "Hillock." The main room was 10 feet by 12 feet, flanked by two 4.5-foot by 7-foot alcoves, one a bedroom and the other a kitchen.

They built shelves, a writing table for each of them, and storage for Sara's painting supplies, "which I hope to use soon!" Not surprisingly, they constructed a safe in the floor, accessible by a trap door, to hide valuables and keep food cold and protected from rodents. A portrait of Henry Wadsworth Longfellow and woodcuts from the Royal Archeological Institute of London adorned the walls.

They'd by no means abandoned botany and, in their spare time, spent hours roaming the surrounding hills and valleys. The type locality for a

Fig. 40. One page of Sara's February 28, 1887, letter to Mattie, on paper that survived the fire that burned their tent. It includes her sketch of the cabin they built. Photo by author. Original at the UC and Jepson Herbaria Archives, University of California, Berkeley.

new species of California poppy, still named *Eschscholzia lemmonii*, was labeled "Lemmons Ranch." They even managed to send seeds of the yellow-flowered bush poppy, *Dendromecon rigida*, a plant often found in newly burned areas, to the Royal Botanic Gardens at Kew. Another new species was the low widespread Lemmon's mustard, *Thelypodium lemmonii*. One specimen collected by Sara and JG still resides at the University of Notre Dame Herbarium.

They expanded their range by using JG's pension to buy two horses for fifty dollars and sixty dollars. Lady-Gray and Johnny-Roan proved "to be very steady—sound & safe—good roaders & if we wish to sell them later, will be able to get the money back & perhaps a little more besides the comfort of having them to ride about." The Steeles provided a double harness and wagon for $20 on trust. Once JG had replaced the brake, he and Sara repainted the conveyance black with red striping. Now the wagon looked so elegant they decided it needed protection from the elements: "Today we rode to a ravine, cut some good oak limbs & they are to be set on west side of house—posts—then we stretch over them some strong unbleached cloth for a wagon shed, 11 × 9 ft. Under this the *newly* painted wagon is to be kept from sun & rain."

They also arranged to buy four hundred oak fence posts for five cents apiece and paid a man to dig two-foot-deep postholes and string the wire for the same price. In addition, they had twelve acres plowed so that it could lie fallow for a year before being planted with wheat.

No question they had improved their 160 acres. The entire project was monumental, and their pleasure in it was mixed:

In time we hope it will turn us something, for all the hard work, worry & anxiety we've had over it. Much enjoyment too has come in fixing up & in air castles, but we are both worked down & I am ill nearly all the time if I overdo. A pain in right side all the time and in small of back on left side. The culmination is neuralgic headache & vomiting. I have been more troubled this way ever since we were in New Orleans.

The atmosphere is balmy & beautiful & I hope when we get rested from the hard work & anxiety, we shall enjoy it.

As worn out as they were, being on the homestead had another advantage: "You may fancy this is no place for any but old clothes. . . . It is lovely and restful to not be obliged to dress for Society for a while at least. We love nature & the freedom it gives us."

17

"Life, to Me, Seems Sweeter Each Year"

Oakland and Mexico, 1887–88

⇢ BY THE TIME THE grasses had once again turned golden and powder-dry on the Hillock, Sara and JG had lived the required six months on their claim at Cholame. As winter closed in, they prepared to head back to Oakland where, as Sara reported to Mattie, "I hope to be better than I have been. I am yet under the Doctor's direction. He says the trouble seems to be inactivity of the liver functions & that is playing the mischief with all else."

Even though JG too was "in wretched health with bowel trouble for the past two weeks," he drove the wagon to San Luis Obispo to settle the final steps of owning their claim.

On October 7, Sara exulted to Mattie:

"Proving up" completed, $250 paid & the 160-acre ranch is ours to have and do with it as we please. No fear of some daring wretch jumping our claim in any indefinite absences. Now we can tear up the ground or leave as a summer fallow, and no one can report or dictate to us. We shall wait for civilization to advance towards "Hillock in Cholame" & hope with five years to get $26+ per acre even though we do not improve it farther. But we mean to try the experiment of setting out fruit trees &c do all we can from time to time as opportunity offers.

Their investment hopes rested on the much-anticipated railroad. "This will help us out a little when helpless illness overtakes us," she wrote. The railroad never arrived.

Sara and JG had spent $1,000 on the Hillock and "the whole effort has pushed us closer to the wall, but we will not borrow & now I am going to tell you that we have only $40 to meet all our expenses till the middle of Jan, then we shall have some more pension money. But I guess we shall worry through as we are living very economically."

As part of living economically that fall, they moved the entire herbarium and their furnishings out of storage to the top floor in the Medical College of the Snell Seminary on Clay Street. The rent was only five dollars per month, probably, as JG reported to Mattie, because of "its great height, up three flights of double stairs! No elevator. The climb is terrible for Amabilis, as such labor is for any woman, and I dread the necessary trips very much. We plan to take as few as possible and put many errands together."

Sara brushed off the difficulty of the stairs, and on October 10, she wrote a long and chatty letter to Clara Barton, inviting "Sister Clara and Brother Hubbell" to camp with them for a couple of weeks at Mount Shasta and Yosemite the following summer. She described their new airy digs as "perfection," adding that once they'd made the laborious trek up to their "sky parlor,"

> we overlook the bay & San Francisco on the west & all Oakland &
> its lovely suburban hills on the east. The room is 16 × 45 feet. We
> partition off a bedroom, then curtain off the middle for a reception
> room & beyond three sides are covered with shelves for the Her-
> barium. In the bedroom are two windows, in the reception room a
> large double window & circular one above, in all about 12 feet long
> & two windows in Herbarium.
>
> Above us is an unfinished tower room that I have thoughts about
> as a studio and observatory. We have a good little telescope & we
> may find much pleasure in constructing a light stairway & so be
> able to utilize this observatory. It is large enough for sleeping room
> should you & Doctor wish to hide away with us in Oakland....
>
> "The latch string is always out" to you dear ones wherever we
> might be in the wide or narrow world.

Despite the stairs, JG agreed the space was worth the effort, as he explained to Mattie:

I feel like having a jubilee. You see all this time I could not see a plant or a book that I needed in any discussion, so I gradually omitted discussion or was conscious of constant errors, so had become quite unhappy. Now 300 plant receptacles and three large tables on castors [sic] with microscopes, dissecting tools, and appliances of various sorts surround me, and I am happy again.

Soon we hope to be doing good scientific work, worthy of the name.

And Sara told Clara, "Life, to me, at least, seems sweeter each year, as I understand more & more of its possibilities. I could truly say & desire to live here for a thousand years & this world is a grand & good dwelling place."

Their New Orleans successes—particularly Sara's—had catapulted them onto the Bay Area's social A-list. One downside of being back in civilization was the expensive requirement to dress for "Society," so Sara was especially grateful when Mattie sent a dress pattern. Sara had it made up in a brown material with a strip of brown velvet on the hem and underside of the sleeves: It was "neat but not gaudy," and she had a hat and gloves made to match. "My winter outfit is assured. . . . I think it will make a nice serviceable gown & look pretty."

That brown dress trimmed in velvet was no doubt useful as the Lemmons spent the 1887 holiday season in Santa Barbara. They were again wrestling real estate, trying to sell a sixty-four- by twenty-foot plot Sara had bought fourteen years earlier. She'd paid $75, and it was now worth $20 per square foot, or $1,280. They also visited San Diego, where they were entertained "royally" by the Kimball Brothers, a family of fruit and olive growers who'd also become friends at the New Orleans Exposition. The Kimballs had suddenly landed in "the lap of great wealth," reported Sara, by selling much of their land to the Texas & Pacific Railroad. They founded what became National City, tucked along the shores of the bay between what are now San Diego and Chula Vista.

"This is truly a wonderful land & past comprehension," Sara commented wistfully to Mattie. "The tide of unbounded prosperity is moving onward at a rapid pace, and we hope to get near a little of the vitalizing salt spray that comes with the gently lapping waves upon the shore."

They spent a day touring the almost-completed Hotel del Coronado, marveling at what was the world's largest resort at the time. The project took up seven acres with its massive theatre, saltwater pool, billiard room, and dining room—with eighty-foot ceilings—that could seat one thousand people. Even more astounding was that electricity illuminated every room.

Sara was especially intrigued watching the fresco painter who "gave us the cost of frescoing the ladies' parlor & reading room—about $2000."

The company also planned to open an immense museum of natural history—an apparent opportunity that excited both Lemmons but never materialized.

Eighteen eighty-eight would prove to be another year of significant losses for Sara and JG. In November 1887, word had flown through the botanical world that a stroke left Asa Gray bedridden and unable to speak. January 30 brought the sad but not unexpected news of his death. Sereno Watson succeeded Gray as head of the Harvard Herbarium, and Sara wrote him, referring to Dr. Gray as "the father of American Botany": "We feel the deepest sympathy for Mrs. Gray & for you, dear friend. How you will miss his cheery sympathetic presence & work by your side! We are thinking of you all over there—realizing too how much of life has gone out of Harvard. We too shall miss him more than we can express."

She'd have been pleased to know that in 2011 the U.S. Post Office released a stamp in Asa Gray's honor.

By now Sara and JG shared an actual salary from the State Board of Forestry, JG as the official botanist and Sara as artist. They were working extremely hard on their first report, a massive one-hundred-page undertaking with the unwieldy full title of *Forestry Report on the Pines of the Pacific Slope, Particularly Those of California: A New Classification* with *Named Divisions, Groups, Etc. Based upon Plainly Evident Characters, Chiefly of the Fruit of Cone.* It also included twenty-four "artotype illustrations"—photographs JG had made, some in forests and others from "prepared specimens of cones, flowers, leaves, seeds, microscopic cross-sections of leaves, etc. taken in our herbarium [that] have been carefully prepared under the supervision of Mrs. Lemmon."

But on April 10, Mattie sent a telegram: Micajah Sawyer Plummer had quietly passed away at the age of ninety. Although the Lemmons were minutes from leaving for a botanical expedition to Mexico—even the lunch box was packed—Sara dropped everything to write her sister a long letter on black-bordered stationery, filled with reminiscences and appreciation for their father and his

> frosted fingers & toes, his scanty clothes & food, the hard times
> he had in helping his mother who struggled with poverty & the
> heart-breaking ills of Grandfather's stage of drinking & neglect; of
> his enlisting in the war of 1812—later of the bloom of his manhood
> almost worn off by his struggles to help his Brother Moses' widow
> & three children, then late in life his own family responsibilities &
> how he had to start with not only an increasing family but Grand-
> father, Grandmother & Auntie all to look after once they became
> too aged to do for themselves. . . .
>
> How full of resolution and courage he was in those days.

Nor did she hold back her deep appreciation for Mattie's years of parental caregiving: "You have stood by many a trying & hard duty & I for one feel that we all owe much gratitude to you for standing so staunchly by this filial duty."

Even though Sara and JG were always chronically short of money, later that year she paid a notary public one dollar to legally relinquish any right to her father's affairs so that Mattie could have a larger share of the inheritance.

Ever practical, Sara also urged Mattie and George to come to the Golden State, now that they were unencumbered: "Come as emigrants with lunch basket—you can make two small mattresses and have your own blankets & beds & be as cozy & far more comfortable than in the ordinary coach unless you take an expensive sleeper & I'd rather take that money to travel on in Cal."

But for the moment, she was almost too sad to travel: "I feel so unfitted for work that were it not imperative as our bread is depending upon it—"

Despite Sara's grief, she and JG followed through with their travel plans.

Beautiful Piñon Pine

Pinus Parryana.—A Pilgrimage Undertaken for Its Rediscovery.

"What is the next rarest California pine to be studied?"
"Parry's, of San Diego County."
"But that species runs over into Mexico, does it not?"
"Yes, and we will follow after it."

So begins Sara's article in the *San Diego Bee* later in 1888. She went on to describe their May Day arrival by steamer from San Diego to Todos Santos Bay, eighty miles south near Ensenada. From there the International Land and Transportation Company of Mexico furnished a "noble span of sorrel horses," a wagon, and a driver named Jesse to take them to the "pine & lake region" at seven thousand feet high in the San Rafael Mountains of northern Baja.

At noon they stopped "in a very paradise of beauty: broad-crowned oaks shielded the sunlight, hung with festoons of clematis, through which gleamed vistas of meadow, waterfall, flower carpeted slopes, and bars of tropic sheen." It was lunchtime, and in a "twinkling" Sara prepared the "eatables," boiling the coffee on an "improvised stove made from an oil can, prepared by cutting out both ends and kindling a fire within—a most convenient, cleanly and expeditious stove, recommended for campers' uses and tested by the writer for the past years during extensive explorations."

Later she wrote Mattie that they bounced along in the wagon for another eight hours before stopping gratefully to spend the night in the village of Real del Castillo, site of a major gold discovery in 1870. Two years later the village became the capital of Baja California Norte and remained so until Ensenada was named the capital in 1882. Since the room the Lemmons rented came without furnishings, they simply moved their camping outfit inside and slept on the floor.

The second evening they set up camp in a corral, a site Sara described as a "safe and comfortable night in a wild, lonely but beautiful region" and where, she reportedly triumphantly, "we found for the first time, *Pinus Parryana*, growing in its own soil. This was what we had come 500+ miles to get." She went on: "Lemmonia went out before breakfast to plan for taking photographs of it, then we gathered fine specimens of the branches and cones so that I might make good water color sketches on returning."

Now known as the Parry pinyon or Parry's nut pine, *Pinus quadrifolia* is a very rare small pinyon pine, first discovered by their friend Charles Parry during the Mexican land survey of 1848. Finding the plant in a new location was an important addition to the Lemmons' exhaustive State Board of Forestry report that would be published in November.

Tantalizingly, in that report, JG also referred to the life-size, larger and more elaborate watercolor paintings by Sara that couldn't be chromolithographed "owing to the much greater expense of such works; but they will be perfected in certain details and augmented by other paintings, completing the series of California pines, when, it is hoped, a future report may be accompanied by these finished and most instructive illustrations." So far, those paintings have not been found.

Feeling pressed by impending publication deadlines, Sara and JG hurried back to Ensenada, just in time for the Cinco de Mayo celebrations. It seemed as though every possible official wished to host a reception for them—but Sara was exhausted, and both she and JG felt they needed to return to California to keep working on the report.

Still, neither exhaustion nor deadlines could keep Sara from observing the opportunities offered by such an appealing landscape and concurring with the overall prediction that within a few years immigration from the United States would change the area. The International Land and Transportation Company had bought the upper part of the peninsula and was offering generous terms to potential colonists: no duties on goods or taxes on land for twenty years. "Soil is rich," she told Mattie. "Plenty of water and a delightful climate, balmy & soft, like Santa Barbara."

And even better, thanks to their railroad passes and the land company's sponsorship, their total cost was fourteen dollars in feed for the horses and driver!

Overall, the Lemmons found this part of Mexico to be "a delightful surprise," one that whetted their appetites for a longer expedition. They began planning a trip in which they'd start near the tip of the Baja peninsula at La Paz and travel north, exploring and collecting all along the 850-mile coast back up to Ensenada.

Despite all the pleasures of Baja, Sara's heart ached, thinking of her father's empty chair. She wrote Mattie from San Diego:

> I can see Father now, tipped back there by the oven door, cane in hand, keeping an eye upon affairs in general, having his say about matters, all from lifelong habit of course. Trying, to be sure, but you & George wisely ignored it & made the best of it, knowing it came from age & infirmity & all that seemed like ill-natured criticism, you attributed to age & that was best. He is gone & I shall always revere & love his memory as a good father & heroic ever battling with adverse things.

June 1888 was a bittersweet month for the Plummer family: They'd barely buried Micajah when Mattie's daughter, Martha Everett, also nicknamed Mattie, graduated from Smith College and married a week later. She had met a young man named Charles Elliott St. John her freshman year, and the two had fallen in love. Charles had studied natural history and botany at Harvard under Asa Gray. But after a theology professor convinced him he'd never find a job in the natural sciences, he changed his major and became a Unitarian minister. Young Mattie and Charles would always remain fascinated by botany, an interest that would eventually benefit both Sara and JG.

Sara apologized to the younger Mattie that she and JG had no extra money for wedding gifts but sent heartfelt congratulations.

Rarely did Sara mention her regrets about not being a mother—but in this one instance she confided in her sister that her children would "be a comfort to your declining years. Ill health & 'the fortunes of war' did not allow us to begin life soon enough & our loss will be greater as time wears on, but we make the best out of life that circumstances will permit & are no worse off than though we had each remained single. So we must not make comparisons or complain if there is no one to care for us when age & increasing infirmities come up on us as they will soon & rapidly."

Despite the transcontinental distance, Plummer family ties remained strong, and Charles Crandall Everett, the fourth of Mattie Everett's children, would prove a comfort to Sara and JG in their declining years. In addition, the younger Mattie and her new husband Charles St. John would eventually have a son, Harold St. John. Thanks especially to his parents' interest in natural history, he too would become a noted botanist. Decades later, he and his mother would help solidify Sara's legacy.

But in the meantime, that summer of 1888 was even more than usually busy for Sara and JG as they perfected the forestry report and finished up fieldwork to Mount Shasta to gather more specimens of the seven species of pines in the region. "I went prepared with a short mountain suit—hob-nailed shoes for the ascent up the wonderful Mt Shasta, 4,444 ft, the only mountain in California with glacial rivers flowing down its north side," wrote Sara.

In July both of them again spoke at the annual Chautauqua gathering, where Sara presented two lectures: one titled "A Botanical Study: Consider the Lilies" and one on ethnology, titled "The Indians of the Colorado River." The San Francisco *Evening Post* described the second as "comprehensive, and gracefully delivered in a conversational manner.

Fig. 41. JG's photograph of the knob-cone pine near Shasta from the Lemmons' book *Pines of the Pacific Slope*. In a letter to Clara Barton dated December 17, 1889, JG identified the woman in the photograph as Sara.

She resumed the subject at the Round Table in the afternoon, and the large number present betokened the interest."

That same month the forestry board held an exhibit of the Lemmons' pine research, and the San Francisco *Morning Call* on July 26 mentioned the "elegant watercolor paintings by Mrs. Lemmon, showing with almost microscopic fidelity the structure and general characteristics of the subject."

In October Sara told Mattie that the forestry report absorbed all of JG's focus—and therefore hers:

> I am attending to all outside matters and relieving him and leaving him undisturbed & with the fewest interruptions till he can see his way out. It is a carefully written report causing close study & work upon his notes & original investigations in the field & we expect it to meet with great favor. It is to be illustrated from photographs, and all my spare time now I devote to retouching the negatives & photographs to get them ready for the work. Then comes proof-reading & that may for a time take us to our Capital City, Sacramento, where the State printing is done.

The State Board of Forestry's shared salary—$150 per month—had supported the couple for eight months and was the most money they'd brought in since being together. But with no notice, later that month the state legislature suddenly informed the Lemmons that the appropriated funding for their salary was exhausted, and they would no longer be paid. Worse yet, the board still hadn't paid for all the hard work California exhibitors had poured into the New Orleans World's Fair.

That winter, politics dominated many a dinner table conversation. On Tuesday, November 6, 1888, Republican Benjamin Harrison—the grandson of America's ninth president, William Henry Harrison—was elected as the nation's twenty-third president after winning the electoral college vote. Grover Cleveland, the incumbent, lost even though he had won the popular vote by ninety thousand votes, or 0.08 percent.

In other political news, according to JG in an irritable letter to Clara, Democrats "and unrepentant rebels" filled other elected slots, a situation the Union veteran and Andersonville Prison survivor found "most galling to me. My cheek burned with shame and I felt revengeful."

He and Sara were still proud Republicans and worried that Prohibition was distracting to the party: "What we, personally, Amabilis and I, believe to be right is to fight the demon of drunkenness within the lines of Republicanism. This grand old party is large enough if all pull together to destroy the saloons and blot out other dens of infamy."

18

"The Narrowest Escape from Instant Death"

Oakland, 1888–91

→ SARA WAS JUSTIFIABLY INCENSED about losing their salaries from the State Board of Forestry, but she was even more outraged about not being paid for their seven grueling months at the New Orleans Exposition three years earlier. Shortly after Christmas the couple took up temporary residence—and battle—in Sacramento.

As always, Sara kept the household running:

> We were fortunate in securing a room, a front parlor comfortably furnished for $3 per week, meals where we please, and we please to take them in our own room. On a kerosene stove I can cook a good steak—we brought coffee, tea, broma, condensed milk, butter, crackers, bread (graham), mince pie & apple cake, raisins, nuts & apples. We engage a quart of new milk & every night I make mush of rolled oats, & with milk we eat just as at home.
>
> I do all the errands, write letters & attend to all outside work to leave L entirely free on proof work.

They both spent many hours at the State Printing Office checking and revising final proofs before the presses ran with the five thousand copies of *Pines of the Pacific Slope*. Five hundred would be theirs to distribute to their customers, family, and friends.

The second reason to be in Sacramento was to attend sessions of the state legislature, which was wrangling over the forestry board's lack of payment to exposition contractors. Each day the couple walked to the legislature buildings where, as Sara reported to Mattie, they met "Senators, assemblymen & their wives, lobbyists and all sorts of peculiar people,"

> but it is important that we watch & work with Legislators in both houses till we learn the fate of the bills. We hope this will be sometime during the next three weeks, else there will be no hope for their passage. The session is only convened once in 2 years and then for 60 days for the immense amount of work to be done. It is interesting to watch their work—something new for me, at least. Our Forestry Board has asked for $110,000—that seems like a large sum. The State Mineralogist has asked for $150,000! The session of both houses, i.e. Senate & Assembly, cost the State at the rate of $12½ every minute of the hours in session. It opens at 10 A.M., recess from 12:30 till 2 P.M., then three hours for the P.M. Soon they'll have night sessions—and there will be tremendous hurry "snowing in" of late bills amid tempestuous roaring of oratory.
>
> We hope to be through before that cataclysm.

At last, on March 21, 1889, the *San Francisco Chronicle* ran a welcome headline: "Mrs. Lemmon Vindicated." Under it was a letter from Sara thanking both branches of the legislature and the governor "for their almost unanimous indorsement [*sic*] of the justice of her claim for $1000, compensation for services for seven months at said exposition."

As JG exulted to Mattie: "Score No. 1 to the Lemmons!" It was no small victory: $1,000 would equal nearly $28,000 in 2020 currency!

Prying actual cash out of the forestry board would prove more difficult.

But by July Sara told Mattie they were almost out of debt, and "the prospect is brighter this season than since we have walked & worked side by side." She even sent ten dollars to her first grandnephew, young Mattie's infant son, "that robust and promising Everett St. John." Her instructions were to invest the money and then give him the principal and interest when "he steps out into the world for himself."

In fall the Lemmons spent several weeks in Oregon and Washington so that JG could photograph the local cedars—although he was thinner than ever and Sara was again ill with gastritis. All she could consume each day was a quart of buttermilk with a side of Carlsbad Sprudel Salz, a laxative for "catarrh of the stomach." She felt better out in the field than cooped up in the house, but added "I look like the last rose of summer with the petals all gone."

In a rare overt reference to her tendency toward depression, she added in an October 7 letter to Mattie, "My other trouble, introversion, is better. So I hope to stem the tide."

Through it all, Sara kept after Mattie to come to California. Plots of land, 50 feet by 150 feet, were available within a mile of their herbarium for $150. Conveniently, they were also adjacent to the just-established Leland Stanford Jr. University, funded to the tune of $15 million, by ex-governor Stanford in memory of his son.

"How I wish the boys could be put there! You have no idea what a stupendous scheme it is!" Sara urged. "And if you wish to go shares, I will invest with you."

As the winter of 1889 shut down the growing season, the Lemmons shifted to their usual indoor activities in their Oakland home and herbarium. Each day Sara sketched from dried specimens, and they worked up their field notes, wrote books and articles, studied new species, and tried to support themselves shipping out cones, seeds, and specimens. The household budget tightened as the promised funds from the forestry board still failed to appear.

On December 17, JG wrote Clara about a recent fright they'd had: a fire in the first-floor stairway that blocked their only escape from the third floor. Fortunately, by chance the Oakland fire department happened to be nearby and rushed to extinguish the flames. "But," JG wrote, "the thought that our herbaria, the collections of 23 years, may be destroyed in an hour quite takes my breath away at times. We *must* get down into a better place next winter *sure*, yet—I dread the labor and expense of moving."

By the following year, 1890, it was still like death to Sara Lemmon to be idle, even at age fifty-four. That same year was the founding of the Pacific

Fig. 42. Sara Allen Plummer Lemmon in 1889. Photo by author. Original at the
UC and Jepson Herbaria Archives, University of California, Berkeley.

Coast Women's Press Association, in which she was a director and an auditor. She also belonged to both the California Press Women and the Woman's Christian Temperance Union and was heavily involved with the suffrage movement—she was referred to as one of the movement's most effective speakers.

In 1890 the *Encyclopedia Britannica* also defined her as an "eminent botanist."

At around this time, Sara mentioned writing "Marine Algae of the Western Shore," which seems to have either been lost or never finished.

That same year Edward Greene named Plummer's mariposa (*Calochortus plummerae*), a new species of lily, in honor of her botanical work. Greene had been curator of the herbarium at the California Academy of Sciences but by now was the first professor of botany at the University of California, Berkeley. Endemic to Southern California, Plummer's mariposa is a small but showy tuliplike flower with petals that vary from lavender to pink to white. It prefers dry rocky places where it's a host plant for the orange tortrix moth.

In April, Ralph O. Bates, a fellow survivor of the Andersonville Prison, stopped by the Lemmon Herbarium. He was traveling around the country giving talks about his Civil War experiences, and Sara described JG and Bates "having heavy exchanges of experiences, breaking down everyone in hearing & themselves overcome with agitation at the 1st meeting. Now they meet & squeeze hands & look into each other's faces as only people with such mutual experiences can."

No doubt the highlight that spring was Mattie's long-anticipated visit, although her husband, George, stayed home to manage the farm. The sisters kept so busy during Mattie's visit that the sole surviving evidence from the trip is Sara's later comment about a popular nautical novel that her sister had left behind: William Clark Russell's *The Wreck of the Grosvenor*. "It was my first sea story," she wrote. "And we found it most exciting."

Entertaining Mattie on top of all her normal responsibilities wore Sara out, but after some rest, JG reported to Mattie in June that she was "not nearly so nervous as formerly, does not have a covering on her head at night, a precaution against cold that has been necessary for many years, and in many other ways is as hearty as I remember for many years—not since we used to go pic-nicing [*sic*] down among the Apaches." His own health was ever more fragile, and he told Clara Barton his voice was so weak he had to give up public speaking, and that, tragically, he could no longer sing.

In August JG collapsed trying to finish the next installment of the forestry report. Sara covered for him, writing the chairman of the State Board of Forestry: "Four days ago, as my husband was writing up the last species, he was obliged to lay down the pen, overcome by illness. A physician and nurse were summoned and he has not been able to sit up an hour since. I am busy copying his revision as carefully and rapidly as possible that it may be in clear, plain & good condition for the printer. The illustrations are about completed."

Two months later she wrote Mattie now that "the awful Biennial Forestry Report is at last in the hands of the State Printer," they could escape to the Sierras:

> We are within a few rods of a beautiful lake, about 1.8 miles across, hemmed in by high mountains that are snow-capped this year from the immense snows of last winter. These mountainsides are covered with dense forests of Pinus murrayana, and we are in a lovely pine grove, our tent, table, and fire surrounded by about a dozen fine ones. The fire-range is made with a long trench dug about a foot deep, stones on the side & an old stove hearth for top. Lots of refuse wood all about us & we can cook hot cakes, fry ham, or boil coffee or tea, roast potatoes & have a torching open camp fire at night.

Sara was still battling gastritis so severe it made her lips, tongue, and mucous membranes sore. A nearby farm provided unlimited dairy goods, so she lived almost entirely on hot sterilized milk enhanced with egg, cream—and port.

October brings chilly nights at eight thousand feet, so she pinned a shawl around her head, and wrote:

> Our sleeping arrangements are quite comfortable. First a bunk of boards about 18 inches deep, 2/3rds full of pine boughs, then a tick stuffed with soft new hay, two comforts, two blankets, lap robe, two feather pillows & your fur-lined cloak for emergency. Then I fill a rubber bag with hot water every night—no cold feet. . . .
> We shall hate to break up the coziness, but cold nights increasing in severity will soon drive us out, we fear.

The Lemmons were camped at the same site they'd shared four years earlier with "Sister Clara" and "Brother Hubbell" and still hung their towels on the wooden pegs that Julian drove into the "kitchen trees." Each evening Sara arranged wildflowers on the table in honor of Clara. Although JG's singing voice and the little silver flute were now silenced, they could almost hear echoes of "Nicodemus" whispering among the pines.

After several more weeks in the mountains, the temperatures dropped, and ice rimmed the water bucket each morning. Reluctantly, the couple headed back to Oakland where Sara began the work to help establish American Red Cross chapters in Oakland and San Francisco.

On December 12, 1890, the State Floral Society voted to make the golden poppy California's state flower. To members of the society, it was an obvious choice: Given enough winter rain, each year the flowers blanketed whole swaths of hillsides in blazing orange. All that was needed was for the legislature to make the choice official. Sara started campaigning for it, unaware she was in for a ten-year fight.

In general, Sara and JG's lives were busy but going smoothly—until a fateful holiday shopping trip. JG told Mattie that Sara

> was looking at goods in a Broadway store and, in order to see articles on the upper shelf, she slowly receded backward, not noticing an open stairway.

I screamed "Amabilis!" but too late.

When I flew down after her and raised her head from the cement floor, I was surprised to hear her speak. The only wonder is that she escaped with her life.

The fall bruised Sara's right side and lumbar region so severely she had to be wrapped in a "rubber plaster" and confined to bed.

The accident became an unwelcome opportunity for Sara to experience firsthand the horrific hospital conditions that were typical in late nineteenth-century western America. Instead of being locked away, medicines were stacked on patients' bedside tables where anyone could help themselves. One physician admitted he had trouble distinguishing medical staff from patients, since patients were so accustomed to nursing each other. Each day Sara saw how ill-prepared and poorly trained most hospital nurses were.

Luckily, once she was released from the hospital, Sara could afford private nursing at home. Two good friends stepped in to help: Ida Forsyth, a fellow New Englander and graduate of the Blockley Training School for Nurses at Philadelphia General Hospital, and Elise Mohl, another trained nurse. Thanks to the three of them, hospital care on the Pacific Slope was about to improve.

As Sara's injuries healed, she and Ida began plotting, combining their Manhattan and Philadelphia medical experiences to develop a training school for nurses. Nine months after the accident, in the autumn of 1891, the two of them met with Dr. John Hopkinson Healy, the superintendent physician of the City and County Hospital of San Francisco, the forerunner of today's Zuckerburg San Francisco General Hospital and Trauma Center. Although Healy supported their plan for a nursing school, the California Board of Health balked at the impropriety of female nurses in a hospital of mixed patients. Besides, the all-male board members were reluctant to remove "the old-time nurses, who though ignorant and inefficient, and some of them often intoxicated, were, nevertheless, men of influence." Sara and Ida persisted, shaming the board by citing the success of similar training schools in the East.

The board relented.

The next obstacle was space, but the two determined women overcame the "loud protests of certain employees"—most likely those old-time nurses and often intoxicated men of influence—to convince hospital authorities to reassign some rooms. Equipment was nonexistent, so Sara and Ida solicited—and received—donations of money, furniture, and bedding from San Francisco residents and friends.

A year later, as reported in an 1898 *Nursing World* magazine article, the San Francisco Training School for Nurses, the first of its kind on the West Coast, became a reality. Miss Ida M. Forsyth took the helm as principal, with Miss Elise K. Mohl as her assistant. Both were also instructors.

The first five nurses, who graduated in 1892, were followed by an increasing number each year. During the 1898 graduation ceremony "frequent references were made to the zeal, hard work and efficiency displayed by the founder of the training school, Mrs. Lemmon, who was invited to be present, and to whom was assigned the privilege of awarding the regulation medals to the graduates."

Oddly, there is no mention of this considerable accomplishment in any of Sara's surviving correspondence.

In the meantime JG was hard at work on the third forestry report. He and John Muir were having a lively correspondence about the phylogenetic home of spruces and which trees belonged in the genus *Abies* and which should be categorized as *Picea*. JG wrote that he appreciated Muir's "frank criticism." He pointed out that "of course botanists don't make species, genera, etc. They exist in Nature, and many of them being more or less concealed by morphology, only the sharp-sighted student has from time to time detected the distinguishing characters."

The report was no small undertaking. On New Year's Day, 1892, Sara wrote a thank-you note to their benefactor Colonel Charles F. Crocker, still the vice president of the Southern Pacific Company, on behalf of JG, who was "temporarily ill with *la grippe*" (the flu). She told him they were working on a three-hundred-page, amply illustrated volume of the evergreen cone-bearers of Northwest America. To Mattie she explained they wanted it to be a work for anyone, not just botanists, interested in plants. The "fine illustrations" would help make it a success for the masses.

The most galling aspect of all their hard work was that, once again, the forestry board had reneged on its promise to pay contractors. In February, as Sara still limped from her fall, she and JG were back in Sacramento, prepared to stay for weeks if necessary, having found a room for two dollars per week one block from the capitol building. She explained to Mattie that this time they were fighting the forestry board "for $2,100+ for fourteen months of work in back pay, plus two months of salary in arrears."

On the morning of February 2, dressed in their Sunday best, Sara and JG were at the state capitol to testify before the Committee of Investigations on State Commissions. That afternoon Sara paused in an alcove of the "lovely state Library, well-warmed and lighted" to scribble a hasty note to Mattie. JG wondered how she could possibly write in the midst of such tension and chaos.

"This is the way I settle my roiled feelings and get calm for the fray," she told Mattie. "If I am called upon tonight, I hope to be calm and undaunted and tell all I know about matters pertaining to our relations with the Forestry work. We have to face some sly and tricky politicians, but plain unvarnished truth usually wins in a just cause."

She added, "JG says we are earning our money twice, and so say I, but it has to be done, and then this kind of experience has its lesson for gaining composure & extra calmness before judicial bodies that is worth something."

The evening session before the committee required their composure and calmness until 11 p.m. that night.

At last, on June 13, 1891, Sara announced to Mattie, "Victory! Victory!! Our claim in full of $2,600+ has been allowed by the State Board of Examiners." The intense intervening months had included more protests, an embargo, and an appeal to secretary of state Edwin Waite, and concluded with assurances that the current board members would be removed and their actions deemed illegal.

"So you see, Mattie, we are at last recognized & solid on the side of power. This will lift our grinding heavy debts with 8 percent interest accruing in all these 2½ years of hard work, deprivation, doctors' bills, sickness & worry."

Sara couldn't celebrate as much as she'd have liked since "a return of *la grippe* has kept me in bed for three days. . . . I am raising a lot of mucous & coughing badly, weak as I can be." But she still sent money to Sadie, Mattie's pregnant older daughter.

She was also at work on yet another idea: a fruit colony for women, "and, Mattie, I want you in it with me. . . . I know of four ladies who will each take at least 12 acres & in five years it will be in good bearing & paid for. There are fine lands all up this way—the Capay Valley all in all the best. I think six or eight women could manage 100 acres." She had already approached the mothers-in-law of both her nieces, Mrs. St. John and Mrs. Humphreys. (The Capay Valley is still known for fertile fields and farm-to-fork festivals.)

The herbarium's continuing struggles to be paid by the forestry board would prevent her idea from reaching fruition.

19

"Sell Everything and Move to California!"

Oakland and Chicago, 1891–93

→ BY FALL OF THAT year, 1891, there was still no payment from the forestry board.

Sara and JG were fed up. In August Sara wrote Mattie, "We are not in any way connected with the State Board of Forestry, nor are we earning a dollar from any source. Our affairs are not yet settled, and we are still borrowing to eke out the necessities, all the while living in hopes of a settlement soon. . . . We do not again propose to work for any of these Government people except to be well paid & and arrangements fully made & recorded *in ink.*" Still, she remained upbeat, adding, "At present we are, as you see, in the hobbles, but by no means cast down or in despair."

One reason for optimism was her improved health: "I have tanned and gained in flesh from the recent camping & roughing it. You would hardly know me. . . . I have drank milk & buttermilk whenever I could get it, eaten lots of fruit, walked up mountain trails miles at a time, thereby toughening the muscles & weakening shoe leather."

Two weeks later, on September 3, her fifty-fifth birthday, Sara wrote Mattie asking if Sadie's baby had arrived yet, adding, "Yesterday we did not have a dollar in our pockets."

That same day she wrote to California's attorney general William Henry Harrison Hart: "Will you kindly relieve my anxiety of mind by stating how our case stands and why is that members of the Board of Forestry Commission, so recreant to justice & their sacred pledge of office, cannot be brought to their honest duty?"

Although Mattie was sympathetic about her sister's financial woes, in December she had to deal with a far worse tragedy of her own: Her youngest son, Georgie, died, apparently from ingesting some kind of poison. He was seventeen.

"We are thinking so hard of you & George that it seems as though there must be some comfort come to you," Sara wrote, while urging Mattie to come grieve in California.

JG added, "Remembering the bright boy Georgie as he ran about the fields collecting flowers for me or as he discussed matters of business so intelligently with his parents. . . . I echo the invitation, dear friends, of Amabilis for you soon to shape your affairs as best you may and emigrate to this coast—not that the dread Destroyer does not here invade—but because his insidious predecessors, pain, sickness & financial cares are less frequent or more easily overcome than in your over-populated East."

Steadfast New Englanders that they were, Mattie and George Everett remained in Dover.

Sara started the new year, 1892, with a note to Colonel Crocker, thanking him for renewing their railroad passes yet again and telling him about their "fully illustrated" three-hundred-page book of the cone-bearers. In March the first edition of the *Hand-Book of West-American Cone-Bearers: Approved English Names with Brief Popular Descriptions of the Cone-Bearing Trees of the Pacific Slope North of Mexico and West of Rocky Mountains* was published, followed by a second edition in April.

In May the Lemmons headed back to the Southwest and to Arizona's Chiricahua Mountains and Camp Rucker, gathering notes and photographs for articles they hoped to sell to magazines.

On May 26 Sara sent Sadie a postcard from Fort Bowie, saying, "We are revisiting the old historic scenes so full of sad pictures in the past when the early emigrants were often attacked by the wily Apache." In another postcard, in June from Juarez, Mexico, Sara told Mattie they'd gathered souvenirs and taken pictures of the tunnel where they'd hidden from Juh's Apaches with the "queer old hermit."

Years later Sara wrote a note to a Pauline Stafford in Bonita Canyon, remembering how Pauline had sheltered JG and Sara in her "crowded but hospitable little house" for a night on their 1892 trip. That little Stafford house is now part of the Faraway Ranch in Chiricahua National Monument.

One of Pauline's Bonita Canyon neighbors was a local rancher named Louis Prue. In a draft article titled "Some Adventures in Apacheland," JG described hiring Prue to drive them in his two-horse open wagon to Camp Rucker. The wind blew so hard that morning all three of them wrapped their heads in scarves as protection from the stinging dust.

Hours later they reached "the broad Sulphur Valley, so rich and beautiful with grasses and flowers previously, a sad view was presented. No grass, no flowers, no streams, only a dry desolated plain with skeletons of animals scattered over it—due to the rapacity and ignorance of stockmen from the East who grazed it three years in succession, with too many thousands of cattle."

In the distance they spotted a lone rider, and as he drew near, they could see he was dark-skinned—and well-armed. Within minutes, he rode up to Sara's side of the wagon.

Suddenly, as she wrote later to Mattie, she was face-to-face with one of those "wily Apaches" with his "2 cartridge belts, 2 live shooters & a short rifle—and we on the plains with a driver unarmed. This circumstance may have saved us."

JG added more detail:

He had long, straight, black hair, an ugly, swarthy, grimy face with deep furrows, high cheek bones and keen black eyes. Approaching to within a few feet, he halted, turned half-round in the saddle, raised his cut-off gun and scanned our faces quickly, back and forth, three or four times while we, frozen to our seat, our hearts beating a tattoo, and with bated breath, gazed into his piercing eyes and resigned ourselves to fate.

Suddenly the fierce look left the rider's face, he lowered his gun, crossed it on the saddle before him and inquired in perfect English, "Have you seen any stray animals, on your way this morning?"

"No," we replied in chorus. "We haven't seen anything!—except three hobbled burros down at White's ranch," added Mr. Prue.

His question seemed an idle one framed at the last moment—for he gave no attention to the answer, neither did he utter another word, but exhaling a full breath, he turned his horse, making angle with his first direction and hurried off as if to avoid scrutiny.

Breathing freely again, we whipped up the horses resuming our course—but each in turn furtively looked back to find our interceptor turned about in his saddle and constantly covering us with his weapon until out of rifle-reach.

Arriving at Camp Rucker a few hours later, they discovered a young San Diego family had built a house near Dr. Monroe's cabin. In typical pioneer hospitality, the homesteaders invited the dust-encrusted, weary, and still-nervous trio for dinner—but Sara and JG insisted on seeing the tunnel first:

What stirring memories! "Ah, the old fire-place around which we gathered to hear stories!" exclaimed Amabilis.

"And there are the same antlers," I cried, "decorated with half-a-dozen dilapidated old hats!"

Creeping into the tunnel how we bumped and stumbled along. It seemed lower, longer, narrower and with more elbows than that of our memory, but the experience within; the want of room in the central elbow, the many discomforts of the long confinement, the mental strain in anticipation of attacks, the constant attention to precautions, the anxiety, the belated deliverance—all came vividly back to us.

Sara could only exclaim, "How did we live in here so long—eleven days. I could not stay here an hour now!"

After everyone returned to the house, all the dinner-table stories centered on recent Apache raids and killings. What with those anecdotes and the recurring image of their rifle-bearing accoster, neither the Lemmons nor Louis Prue slept much. By morning all three were so thoroughly spooked, they called off the rest of the Rucker trip in favor of a hasty return to Fort Bowie.

All the way across the grassless valley past the bleaching bovine skel-etons, each of them watched the horizon intently—but encountered no one. Once back in the safety of the fort, they described the interaction to their friend Commander Rafferty, who immediately exclaimed, "That was none other than the Apache Kid!"

Haskay-bay-nay-ntayl, more often known as the Apache Kid, was a member of the San Carlos Apache tribe and had worked as a respected Indian scout, serving in the U.S. Army from 1881 to 1887. In May 1887 he'd been part of an alcohol-fueled celebration that went wrong when two men were killed. The Kid was court-martialed and sentenced to prison. He escaped and, from then on, was on the run. His reputation grew legendary, and he was frequently linked—sometimes with no real evidence—to numerous killings around the Southwest.

Two days earlier, Rafferty told the Lemmons, the Apache Kid had murdered a family near the Mexican line, and the day after stopping the Lemmons and Prue, he'd ambushed a sheepherder in the Whetstone Mountains twelve miles west. Several troops were out hunting for him with the added enticement of a $1,500 reward.

Several weeks later, after Sara and JG arrived home in Oakland, one of the Fort Bowie officers sent them a telegram: "Louis Prue killed Monday, near his ranch, by Apache Kid."

In July Sara told Mattie their forestry claim still languished exactly where it had been a year earlier. She added, "If we get the claim, we mean to send you a present, also Sadie's dear little girl. It is in our hearts all the while, but it seems to me the pressure is daily grinding deeper and the cramping more unbearable."

In late fall of 1892 Sara and JG made a second trip to the Arizona Terri-tory that year, this time to the northern part of the state. George Wharton James's book, *In and around the Grand Canyon*, published in 1900, lists them as November 4–5 guests of the Canyon Hotel at the Peach Springs Trail, by then accessible via the Santa Fe Railroad. They explored much of Peach Springs Wash and Diamond Creek to the confluence with the Colorado River and provided a full plant list, including the description of the cane cholla (then *Opuntia arborescens,* now *Cylindropuntia imbricata*) as "quite large and terrible bushes."

In early December Sara and JG traveled on to Albuquerque, where they spent three days exploring the Sandia Mountains, including Camp Whitcomb at eight thousand feet. At first they stayed at the European Hotel for seventy-five cents a night. But G. W. Meylert, the proprietor of the San Felipe Hotel, was an old friend of JG's and read about the couple's arrival in the paper. He immediately invited them to stay for free at his establishment, which advertised itself as the "Leading Hotel of the Rio Grande Valley" and usually charged two to three dollars per day.

They were back in Oakland by Christmas and, according to JG's letter to Mattie, "at work like beavers, but it is by *choice,* and we are very happy in it. Traveling and climbing, delving and collecting, poring and picturing—so the time glides joyously by."

Time was gliding by for Oakland as well. Stone and brick buildings had sprouted up along all the principal streets, while new residences stretched throughout the surrounding countryside. Even the streets themselves were transformed: "Horse cars are gone, and cables or trolleys are almost everywhere," JG wrote. Even Clay Street, the site of their home and herbarium, was "no longer the still dark lane" it used to be.

Sara added that the town was "being girdled and ribbed with electric cars . . . and great buildings loom high up in the distance as seen from the windows."

Another world's fair beckoned, but fares across the country were pricey. Sara queried the Canadian Pacific Railway, suggesting that in return for articles for the company brochures, she and JG be given free passes. She referred to the couple's many published pieces in *Overland Magazine, San Francisco Chronicle, Pacific Rural Press,* and other publications, reassuring the officials the railway would be acknowledged in every article. She signed her letter as "J.G. Lemmon & wife, botanists and explorers."

The railway rejected her pitch. Sara was so used to success that failing to score the assignment sent her to her bed for the day.

Nevertheless, the Lemmons scraped up funding, and in July they boarded the train bound for the World's Columbian Exposition—more often known as the Chicago World's Fair—with stopover privileges at Tacoma, the Selkirk Mountains, and the Great Dominion Park at Banff. "We have never had a more delightful trip than this last one of 1000 miles in the forests & mountains of British Columbia," JG reported to a friend.

Dreading the "suffocating heat and clamoring crowds of Chicago," they resolved to have a good time anyway. The trip cost eighty dollars, an expenditure so steep they had decided to sleep on camp cots in the apartment of a friend, writer Mary H. Field, three blocks from Jackson Park and the fair itself.

But to their surprise, they loved the fair. The weather was delightful, and "camping" in Mary's sitting room at the Chautauqua House was convivial and fun. They spent many hours in the Forestry Building as well as perusing the "miles on miles of paintings in the Art Gallery." The building with anthropological studies needed a week's examination alone. Fair food was pricey, but Sara packed lunches for them, and they bought nothing on the fairgrounds—except of course popcorn.

Sara had her own responsibilities at the fair. She was listed among women in the sciences as "artist of the California Board of Forestry" at the Women's Pavilion, but more significantly the National Floral Emblem Society of America, which had been loosely organized back in New Orleans during the 1884 World's Fair, was officially launched August 25, 1893, at the Chicago World's Fair. Everyone agreed on the need for state flowers—and a national one—but the choices were so hotly contested that a committee, and yet another organization, were formed the same day to study the topic.

Even though she wasn't at that particular meeting, Sara was appointed chairman of the new California State Floral Society. She and others felt passionately California's emblem should be the iconic golden poppy. Hundreds of poets, including Sharlot Hall, Ina Coolbrith, Charlotte Gilman Stetson, and Bret Harte dedicated verses to the iconic and lovely "Copa de Oro."

In addition to Sara's work with the floral society, on Tuesday, August 29, she also gave a talk on women in science, both "single and mated."

The Lemmons had concocted a complicated travel plan, involving leaving Chicago midfair for side trips to Michigan to visit JG's brother and to Dover to see Mattie. Then all of them would gather back in Chicago for more of the fair after JG's stopover in New York to see about publication of their "big book" on the cone-bearers.

Once again illness intervened. In October JG caught "a heavy cold," which then threatened to turn into pneumonia. He was forced to retreat, alone, back to Oakland—while commissioning Sara to oversee the book arrangements. Her revised itinerary was to continue on to Dover via New York, Washington to see Clara Barton, and Pittsburgh, where she would visit her niece Mattie Everett St. John.

Both Sara and JG were still fiercely determined to persuade Mattie and George to try their fortunes in California. On October 5, Sara wrote Mattie from Chicago:

> In the meantime I just want you, Mattie & George, to earnestly seek to arrange as fast as you can to break up life in that musty old Dover & when I get there we will all bend upon it to pack up and be off as fast as we can. Do not discuss it with everyone who will not lend a hand . . . and in a week we can do all the packing. . . . Make arrangements to sell cows, horses, hens, at certain time put premises into hands of someone to rent—or better to sell—and let nobody or anything deter you from getting away.
>
> We will talk it all over when I come by and large, but do not let everybody pull you down & keep you there.

By now Mattie must have been heartily sick of being hammered at by her older sister, and one can only imagine the intensity of the discussions once Sara arrived. The result? Sara never again mentioned the idea of Mattie moving to California—or at least not for several years.

Sara spent the entire autumn and early winter in the East without her beloved husband and at last boarded the train for Oakland several days before Christmas. The sisters parted on affectionate terms, and Sara wrote a quick postcard to Mattie from Kansas City, thanking her for the picnic basket of food she'd packed, saying, "The lunch holds out splendidly. I think marble cake is the acme of lunch on the road. Everything reminds of the loving thought that prompted the hand."

She sent another postcard Christmas Day, saying she'd be arriving home the next day at 6:15 at the Sixteenth Street Depot in Oakland, "where I hope to find JG as I wired him at Wordsworth today. I have been thinking of our scattered family much & often today."

Sustained by Mattie's marble cake all the way across the continent, Amabilis made it safely home to her Lemmonia in time to see in the New Year.

20

"A Sweet, Sacred Togetherness"

Oakland and Mexico, 1894–98

→ SARA AND JG, ALTHOUGH chronically short of money, were content. Their lives orbited around one another, their work, and their friends. They still hung out with the nurse Ida Forsyth, the artist Adelia Gates, geologist and botanist John Muir, and California's first poet laureate, Ina Coolbrith. Even Caroline Whiting, Sara's longtime friend and boss from her Manhattan teaching days, came to visit. The couple also spent a convivial week camping at Mount Tamalpais with Yosemite promoter James Hutchings and his wife.

In addition, they managed to fit in collecting trips throughout the Sierras and elsewhere in the West and were still hard at work on the big book of cone-bearers while developing a smaller pocket guide.

Their herbarium was a frequent stop on local cultural tours. One visit by the San Francisco Sketch Club was so popular that the forty attendees had to divide up into two separate days.

Sara was also an early and active member of the Pacific Coast Women's Press Association (PCWPA), formed in 1890. Although the organization focused at first on women journalists, its range later expanded to include authors "in good standing in the state." Sara was particularly pleased when the association's official publication, the *Impress*, managed by Charlotte Perkins Gilman, shifted from a monthly format to a weekly that would focus on "literature, science, art, and women's interests in general." Sadly, the journal lasted only twenty weeks, possibly because of Gilman's reputation as a divorcee and a mother who'd relinquished her daughter to her ex-husband.

Fig. 43. Even into their sixties and seventies, Sara and JG made a point of getting out into the forests every year. Photo by author. Original at the UC and Jepson Herbaria Archives, University of California, Berkeley.

By 1893 the PCWPA was a leader in the growing women's movement. Its directors initiated the West Coast Women's Congress Association, a series of gatherings jammed with as many as two thousand women—"as bright, clever, well-read women as you can well meet," Sara wrote Mattie.

She was both flattered and apprehensive when, in 1894, the PCWPA invited her to present a paper on botany: "JG helped me out on my paper: criticized, suggested and worked over it as though his own. He tested my voice & time, &c, then went to the Congress & listened at the farthest part of the big Golden Gate Hall. It was packed with people."

Modestly she reported later, "The paper well received."

Of course there was no payment, but the following month Sara and JG each received ten dollars for a talk at Mills College. A reporter for the *San Francisco Chronicle*, obviously enamored with the couple, described their finances in April 1894: "In their home which is so characteristic of their ideals and aspirations, this model pair reside, and in their life there is a constant succession of research, discovery and attainment. They are very unworldly, and like all true scholars, poor, for the real scholar has no time to earn mere dollars."

On top of their fiscal worries, both were still battling physical woes. JG was tired and thin, and Sara ached from a disagreeable shoulder pain that had been diagnosed as a heart obstruction that blocked circulation and "thus gives pain to nerves." The solution was to "eat but little meat, plenty of eggs, mild fruit; no cucumbers with vinegar, no sweet things—only the sweet that comes in grapes or any fruit." She was also to avoid "any undue excitement, or worry, or rapid excited movements. . . . I hope to get over it by care, but we can never tell when the death blow is really dealt," she told Mattie philosophically.

In December 1894 Sara landed an assignment creating five detailed pen-and-ink botanical illustrations for a school book, the California State Board of Education's *Revised Third Reader*. "The grouping, composition & work must be done with utmost care & skill to fit exact places after the pictures are photographed. . . . I can only work upon them a few hours a day," she told her sister. "In a mercenary way a $30 job, but it is such fine careful work I half fear it will hurt my eyes more than that value. But it is bread & butter, but not 'pot-boiler' work," she added. The pay was later bumped up to $50 for the five illustrations.

"Everyone who saw them pronounced them beautiful," she told her friend Caroline Whiting later.

JG's voice had recovered, and they were both focusing more of their lectures and writings on the vital importance of forest conservation. And as a consequence of their friendship with Clara Barton, Sara was also starting work on a history of the American Red Cross on the Pacific Slope.

As if all those activities weren't enough, Sara was continuing the campaign to make the golden poppy California's state flower. In 1894 alone she'd sent out five hundred circulars and canvassed the entire state, addressing relevant associations—where everyone agreed the poppy should be chosen. Over the holiday while wrapping up the *Revised Third Reader* illustrations, she assembled and submitted all the paperwork for the state flower to Alameda's state senator of the Third District, Republican Guy C. Earl, who would present the proposal as Senate Bill 707 to the state legislature in Sacramento early in the new year.

As of March 1895, the Lemmons still hadn't received any payment from the Board of Forestry, but they were optimistic. They'd again traveled to Sacramento, and after JG's two-hour presentation to the legislature, Sara announced to Mattie that the governor "signed the bill for deficiency in State Board of Forestry expenses"—which included the Lemmons' invoice. "Now we have to wait upon the Board of Forestry, the confounded rascals—I wish they were breaking stones in jail. . . . It seems interminable to us & we venture doubly so to the creditors. If it had not gone through the Legislature, I do not know what we would have done to retrieve ourselves. I guess JG would have sunk under the load & I too."

The same legislature also offered hope to the golden poppy: Both the Senate and the Assembly passed Senate Bill 707 almost unanimously. Yet Governor James Budd, a Democrat, refused to sign it, "thus ignoring the general sentiment and wish of the people, and for personal political reasons, well understood at the time," Sara fumed. Those reasons probably

to help each make seed to sow the earth for the next year.

In the higher regions of the State the flowers come later, after the snows have gone, and one who

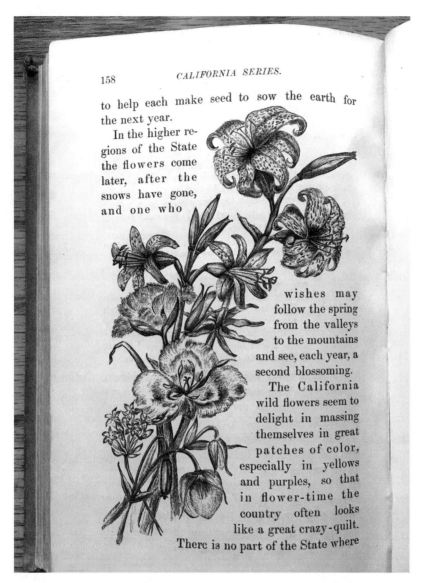

wishes may follow the spring from the valleys to the mountains and see, each year, a second blossoming.

The California wild flowers seem to delight in massing themselves in great patches of color, especially in yellows and purples, so that in flower-time the country often looks like a great crazy-quilt.

There is no part of the State where

Fig. 44. Sara's pen-and-ink illustration of the lily family for the *Revised Third Reader*, published in 1895 by the California Board of Education. Sara described it to her sister: "One group was liliaceous: A Humboldt, a Lady Washington lily, a Calochortus, & Tritelia, all with buds, half-full blown flowers & seed vessels." The other four illustrations in the book were of a yucca and the rose and pea families. Photo by author.

had to do with the bitter gubernatorial campaign in which Republican newspapers had accused Budd of raping a maid—an allegation he called "an infamous falsehood."

Nevertheless, poppy supporters were grateful for any progress. The *San Francisco Call* reported on March 17 that "to Senator Earl was presented a huge bouquet of California poppies, the gift of the ladies of Sacramento in behalf of his efforts to make the Eschscholtzia the State flower."

On top of all their other endeavors, the Lemmons were finally moving from their rented firetrap in the sky to their very own home at what would become their final address.

Clay Street had grown into a bustling but noisy business center, and every day seemed to bring yet another building. Their new address was in Temescal, practically out in the country, on the corner of Telegraph Avenue and Thorne Street. (Two years later, in 1897, Oakland would annex Temescal.)

Although they looked forward to being settled in their new home, "What an everlasting job packing our boxes proves to be!" Sara complained to Mattie in June. "It's not like ordinary house furnishings—and dust has accumulated in many things that are packed away, making it sort of sickening to handle." JG was down with a cold, leaving most of the work to Sara.

But she was excited: A new pump had just been installed, and she declared their water to be soft and delicious. The twelve-foot by eighteen-foot chicken house where she planned to raise Plymouth Rock hens for eggs and for sale was erected and already partly shingled. The plaster on their own house needed two more weeks to dry, so to avoid inhaling any dampness they camped in the yard in their well-used tent.

September 3, 1895, was Sara's fifty-ninth birthday, and she and JG celebrated it quietly—in their new home. That evening she wrote her friend Caroline Whiting: "We have just lighted our evening fire upon the hearth, and the chimney draws smoke up its throat perfectly. We placed our circle of chairs around it for the absent & chatted & read the long evening while the fire glowed & sent many bright sparks up as if to

Fig. 45. Sara and JG's house on Telegraph Avenue, 1897. Photo by author. Original at the UC and Jepson Herbaria Archives, University of California, Berkeley.

illumine the faces of the distant loved ones. I wish you could see, dear friend, how cozy we are beginning to be."

They still lacked curtains and carpets but were content without them to avoid going further into debt: "'Plain living & high thinking' is our motto & we hope always will be. If we could command riches, we should always keep to plain simple ways. There is more life & more to life not to be slaves to fashion & frivolity. One can be so much more useful & happy in the world."

Sara and JG adored their new home, which was an ideal blend for living, working, entertainment, storage, and exhibit space. Sunlight illuminated every room, and the house was spacious enough for guests. As JG wrote to Clara Barton, "We are glad now to return the hospitality so often extended to us—and none more welcome than you and the dear Doctor."

Always a stickler for details, Sara told Caroline Whiting more about the house:

We have a front parlor, bay window—size of room 16 × 16 ft. The middle room is the library with low bookcases to fit each space & here is the little fire-place and red tile cream & brown bordered hearth—size of room 16 × 16 feet with folding doors. Next elevated to broad easy steps is the "Cone room"—12 × 16 ft, 2 windows & a sash door, all spaces with closed closets, shelved, for a part of the Herbarium specimens. This finished in the natural color of the Redwood Sequoia sempervirens (see book). The room is in form like a huge cone—or shell. This is to be decorated with all the different species of cones of W. N Am—a fine object lesson. The room looks like a miniature stage.

Sara often paused in their front hall to admire the view of every room on the ground floor. The hall doubled as a reception area with a cushioned divan. Even the stairway with its broad landing and a "peekaboo" window overlooking the avenue was "a delight to step upon." Upstairs were three lovely sunny bedrooms, one with a stepladder to a sky-lit attic where they could store their trunks.

Fig. 46. The Lemmon Herbarium, showing Sara and JG in their workspace. Photo by author. Original at the UC and Jepson Herbaria Archives, University of California, Berkeley.

Shortly after they moved in, she told Caroline, a "very puzzling" package arrived. Sara unwrapped it to find a full dinner set of seventy-five pieces: knives, forks, dessert and table spoons, all of inlaid sterling silver! The gift was accompanied by a two-page poem that only added to the mystery. Who could have sent it? "Someone who knows enough about our domestic, culinary affairs to learn this lack," she mused.

Was it Ina Coolbrith? Or maybe Mary Field? Both were poets—but so were many people in those days, including Sara herself.

If the Lemmons learned the identity of the generous gift giver, they never mentioned it in their letters. In any case, the silverware was very welcome.

The *Revised Third Reader* had been published that summer, and with her customary modesty, Sara told Caroline the drawings "surely look well. That is all I venture to say."

Their pocket guide, *Handbook of West-American Cone-Bearers*, had been delayed at the printers but was also finally published. Enclosing a copy to Caroline, she said, "It is much in a small compass and a convenient companion for student and tourist who loves trees enough to desire to recognize & learn of them."

In October 1895 the couple traveled to Mexico. None of Sara's letters from the trip have survived, but an article in the October 27 *Mexican Herald* described their two-month travels, announcing, "It is the plan of Prof. and Mrs. Lemmon to make several trips to this country, spending their time in the immense work of thoroughly classifying the trees of the districts they will visit, as work which has never been done here to any great degree."

Having enjoyed their stay with Mary Field in Chicago, they joined her again, this time in Mixcoac, near Mexico City. Here they explored Lake Chapala and Guadalajara while Sara illustrated and JG photographed the forest trees.

In December they returned home to a surprise party that celebrated both the new house and their fifteenth wedding anniversary. (At least, it was a surprise to JG—Sara had been let in on the secret four days earlier.) "It was a rouser, no mistake, and warmed up every room from cellar to garret," JG reported to Mattie. "Congratulations, encouraging remarks, gifts, etc. flowed in upon us. . . . a very flattering affair."

By now, they were well-known. Years later botanist and professor Willis Jepson recalled, "In the 80s and 90s if you asked any chance person in California the name of a botanist, if he knew the name of a botanist, the chances are very greatly [sic] that he would have known the name of Lemmon and only Lemmon."

During the next few years they lectured more and more. A March 22, 1897, Pacific Coast Women's Press Association program lists a talk by Sara ("Camp Life and Study in the Woods") as well as one by JG ("The Forest Endowment of the Pacific Slope"). In addition, JG kept busy in the Neighborhood Club, the Good Humor Club, and the Telegraph Avenue Improvement Club, and spent eight years as the chaplain for the GAR's Admiral Porter post. He also puttered about the yard, building seven feeding troughs for the chickens, plus the "heavy task I've taken upon to rid the yard of a certain vile underground, creeping weed." He told Ida Forsyth he visited each of their growing trees every day.

Although Sara and JG were now in their sixties, they were still eager to explore. In September 1897 Sara wrote Mattie about their three-week collecting trip in Yosemite—fifty miles from the nearest post office without any news from the outside world.

Nearly forty years earlier Congress had granted the Yosemite Valley and the Mariposa Big Tree Grove to the State of California—after burning out the Ahwahneechee tribe who'd lived in the Valley for centuries. By 1890 a mix of political infighting, insufficient funding, and unwise management threatened to destroy the once-pristine area, leaving it vulnerable to the ravages of lumbermen and domestic sheep, which the Lemmons' friend John Muir referred to as "hoofed locusts." In 1890 another act was passed; this one assigned a "Guardian" and two troops of cavalry during the summer months to oversee and protect the park.

Sara and JG's expedition, seven years later, was magical:

Something of grand and brimful of interest in Nature, in rugged mountain scenery, immense belted forests, long rivers, steep cañons, roaring tumbling waterfalls, bears, coyotes, deer. We've been up mountains 13,000 ft, have been caught in two or three heavy thunder tempests & hail storms—a tree struck by lightning near us & set on fire. Night has overtaken us on a long trail & with our horses unsaddled & tethered, we've built a big camp-fire & lain down on our saddle blankets with saddles for pillows with a small cold supper & breakfast, then back to camp triumphant—for we've had a successful trip.

They weren't alone the entire time: The former Guardian of Yosemite Park, Mr. Galen Clark, joined them with his saddle-pack horse, even to the top of Mount Dana. He was eighty-three years old and still pitched in to help repair their wagon when the forward axle broke.

They went on to Hetch-Hetchy Cañon and to the grove of giant sequoias.

Sara and JG saw the soldiers who "guard this Park of thousands of acres that no sheep men, hunters, etc. encroach upon it or fires rage therein." It wasn't until 1903, six years later, that President Teddy Roosevelt accompanied John Muir for a night in Yosemite's backcountry, an occasion both would always remember. Yosemite became part of the National Parks system in 1916.

Throughout their own trip, Sara continued to paint. She described her illustration of the Incense Cedar to Mattie: "It is a beautiful study but full of detail and requires much patience amid many interruptions and camp duties." Sadly, that painting is among the many that have been lost.

Those camp duties still included cooking, which Sara turned into yet another creative endeavor, sizzling up "a mess of delicious trout" and baking biscuits, hot cakes, gingerbread, and pies over the campfire in a frying pan covered with an old pot lid.

"What do you think of that?" she asked Mattie, adding defiantly, "It will end, this camp life, in my becoming a very careless housekeeper

& haphazard cook, according to the good housekeeper's standard. But what of it?"

As one of those hyperorganized women who sees what needs to be done—and how—Sara apparently also had trouble saying no. The Philippine-American war was heating up, and she predicted the thousands "of men who have gone to the Philippines will return this way and will not be as vigorous as when they started out—and that means Red Cross work for months to come after war ceases." By 1898 she was soliciting funds for the organization on her own "Correspondence Secretary for the Oakland Red Cross Society" stationery, appropriately printed in red ink.

Her new responsibilities included yet more presentations, these to inform "citizens what we have been doing with the funds—$10,000+ and material that they have so generously donated for the Red Cross work." She and JG also hosted visiting Red Cross dignitaries, including an old friend, Connecticut's Judge Joseph Sheldon, a fellow Unitarian and an advocate of both women's suffrage and temperance.

By now Sara's resolution to stop badgering Mattie to move West had wilted, but despite their disagreement on the topic, that stout-hearted sister had ventured west for another visit sometime around 1898. One photograph of the two of them in Yosemite has survived.

In 1898, to both Sara and JG's delight, Mattie's fourth child, Charley, came to explore his options in Oakland—and stayed. The following year Mattie Everett St. John would deliver another reason for Mattie to remain firmly in the East: twin sons, Sara's grandnephews. Born at seven months, they weighed only two pounds each, which was worrisome to all.

In November 1898 Sara took a trip to Pomona to visit her Aunt Salome, and to Los Angeles for some Red Cross business. She'd been traveling on her own more, but JG still missed her, writing: "As I returned home, the house never seemed so empty before. You would think that I had got used to your absence of late, but no, the knowledge now that you were to be gone a long time constantly weighs upon me. Even in my dreams last night you were calling out 'Lemmonia' in the usual tones of endearment."

Fig. 47. The two sisters, Sara Plummer Lemmon and Mattie Plummer Everett, in an undated photograph taken at Yosemite, probably by JG. Photo by author. Original at the UC and Jepson Herbaria Archives, University of California, Berkeley.

November 24, 1898, Thanksgiving Day, was their eighteenth wedding anniversary, and Sara wrote him:

> As I never mean to be away from your side another Thanksgiving, our Wedding Anniversary, while we two, hand in hand trudge down the hill of life together. It will, by sunset today, be forever out of my power to make it a love-day as well as thanks-giving day—at a distance to my most precious friend and dear comrade husband. . . .
>
> I hope that although we are for the time, separated, dear Heart, the life-chain of our love & life together will not have had one link lost or broken . . .

This unique blessing embodies the wish that we may enjoy many, many anniversaries of a Sweet, sacred togetherness.

Mail delivery was unreliable, and JG was waiting for her reply when he wrote, "I hope to find a letter from my best friend, one who shares every waking thought as well as my dreams."

Sara was still in Los Angeles ten days later when he wrote: "Today cold and windy, continuing tonight. Such weather always makes me feel gloomy—and tonight I'm doubly lonesome. If the Light of my Life were here now, the gloom would be dispersed instantly!" He then added, "About bed-making you know I'm peculiar, and you would think so indeed if you could look in upon me and see me placing your pillow in place every night and at the usual time turning toward it and *kissing the place where your head usually lies!*"

No letter has survived detailing Sara's return to Oakland, but it was, no doubt, a joyful reunion of the two best friends.

21

"Wish We Were Out in the Wild Woods"

Oakland and Arizona, 1899–1903

-→ AS 1899 DAWNED, BOTH Sara and JG struggled together
to push the golden poppy bill through the legislature. At Sara's request,
Oakland's assemblyman J. A. Bliss redrafted it into Assembly Bill 229.

Again it easily passed both branches of the legislature, as it had four
years earlier—until being vetoed by then-governor James Budd. On
February 17, 1899, Sara wrote the new governor of California, Henry Tifft
Gage, reminding him that for six years she'd canvassed every important state convention—and not one of them chose a different flower.
"Although it is a measure of sentiment by the people rather than that of
monetary value, yet it is important to cherish a pronounced sentiment
of the people. Great indignation was expressed by the State when it was
learned that a similar bill passed almost unanimously by both branches
of the Legislature during Gov. Budd's administration & he failed for
some reason to place his signature upon the document."

She even tried shaming Gage: "We trust you will have no hesitation in
signing this harmless Bill as many other States have selected and legalized
their Floral Emblems."

So weakened by two months of influenza that he had to write in pencil,
JG added, "It would be humiliating to have our grand California lagging
behind any longer."

Yet the new governor also vetoed the bill, saying flower adoption wasn't
a suitable topic for legislation. Again, Sara blamed political partisanship,
as she explained to Emory Smith, the author of *The Golden Poppy*, published later that year: "The acrimonious fight which was in progress for
the selection of United States Senator was an unlooked-for hindrance.

Assemblyman Bliss, unhappily, was in opposition to Governor Gage's candidate and so failed to get his bill signed by the Executive."

Sara was undeterred, telling Smith: "The bill will be introduced again at the next session of the Legislature, and, if necessary, again and again, until a Governor is found who is broad enough to bury his petty animosities in the interest of the people whose servant he is, and who has good sense enough to encourage innocent sentiment and patriotism."

Governor Gage soon had his hands full with weightier matters than the state's floral emblem. The new century would bring the ship *Australia* to San Francisco Bay—along with a hefty cargo of rats bearing bubonic plague. The disease moved into the crowded tenements of Chinatown. Gage would spend the next several years trying to protect the state's economy—and the pocketbooks of his cronies—by fiercely and falsely denying the pestilence outbreak, despite the refusal by several states to accept any goods that came from California.

Sara and JG were still close to William Lemmon, JG's oldest brother, and visited him in Sierraville as often as their own fragile health allowed. In August 1900 they made the trip, even though Sara said JG was still "very weak and thin and much of the time seems quiet & spiritless & so I know he does not feel as well as we could wish." William was eight years older than JG and, as Sara told Mattie in a long and chatty letter August 20, "he has a very tender feeling towards JG as the baby brother being an infant when he was a big boy. And at the time of the father's death, he took a sort of fatherly charge and was the stay & support of the widowed mother & little family."

By this visit, William was seventy-five, and "still retains his legal and intellectual faculties perfectly and is a great reader." But he was also asthmatic and had to fight hard for every breath. Between that effort and his failing eyesight and hearing, he told Sara and JG that he had "nothing to look forward to and hopes he may not survive another winter." Fortunately, he rallied during their visit.

By then, William and JG's sister, Rebecca Lemmon Olesen, had moved next door to keep house for him. Four years earlier her husband had left her for another woman—taking all $15,000 of the household's money. According to Sara, the betrayal drove Rebecca "insane for a time but she seemed to recover."

Rebecca was fond of her youngest brother, and she'd followed all of JG's career with interest, especially after working with him in Nashville hospitals during the Civil War. During this particular visit, she arranged a reception for Sara and him—inviting four hundred friends, neighbors, and even JG's former students from twenty-five years earlier.

JG entertained the crowd with a botany class, using prepared specimens of all the local conifers. Sara followed up with a presentation on the snow plant—*Sarcodes sanguinea*—an uncommon and strikingly beautiful scarlet wildflower. "It was an exhilarating affair," she reported. "Everybody listened attentively and seemed pleased."

By this time Mattie's son Charley had settled in at the Oakland house with Sara and JG. Sara reported to his mother that he was enjoying his San Francisco job "that would be worth a big salary in years to come." The sedentary office work wouldn't be detrimental, she assured Mattie, because he was retaining vigor by walking the few blocks from Telegraph Avenue to the dock, breathing the brisk Bay air of the ferry ride, then walking again to his office.

She still couldn't stand to be idle and also mentioned "during my odd moments, I wish to study French." She was taking lessons from a friend—who accepted payment in eggs.

The new year—and new century—brought grief to the family. In February, Mattie St. John's sons, the premature twins and Sara's grandnephews, died at nine months old within a week of one another. Years later, their older brother Harold blamed "milk fever"—probably a form of scarlet fever transmitted by milk handled by a dairyman already sick with the disease.

In spring Rebecca Olesen came down with the dreaded grippe. Whether as a side effect of the influenza or the financial stress of her husband's defection, on May 1 word arrived at the Lemmon Herbarium that Rebecca's mind had failed.

Three days later, Sara set out to help, as she told Mattie: "On the 4th I started on the hard, cold R.R. ride over the mountains to Truckee & then by stage 28 miles again over the rib of the high Sierras to Sierraville—the horses floundering and wallowing in the mud & past big snowdrifts. A very fatiguing journey to find Mrs. Olesen almost if not entirely irresponsible. The form of mania was incessant talking, night & day."

Collectively, the family decided Rebecca needed to move somewhere less remote than Webber Lake. Sara helped her sister-in-law into the stagecoach and accompanied her back through the mud and snowdrifts and onto the train all the way to Oakland. "Under protest from the doctors," she and JG then kept the anxious, babbling woman in their house for three days and two nights. They escorted Rebecca before the Examining Board for Insanity and then reluctantly committed her to Stockton State Hospital, hoping her sanity might be restored by the facility's skilled treatment. The institution, also known as the Insane Asylum of California, had just been set up two years earlier by the State Commission in Lunacy.

But several days later, Sara reported that "word came from the Dr. Superintendent of the hospital that the outlook just now is not favorable. Her age, 76, is against her, and this trouble is recurrent."

JG resigned himself to becoming Rebecca's legal guardian to handle her financial affairs. "Everything possible will be done for her," said Sara, who asked that Mattie not mention the event outside the family. She concluded, "It is painfully sad that her light should go out in this way."

It's difficult to overestimate the importance of women's organizations at the turn of the century. The clubs not only offered women a social life and opportunities for higher education in both the arts and the sciences; they were also an avenue for community improvement in labor rights, educational development, and more.

In 1890 Jane Cunningham Croly formed the General Federation of Women's Clubs International (GFWC) as an umbrella organization to coordinate the hundreds of clubs. Six years later, Black women established a parallel movement, the National Association of Colored Women. In Oakland, the California State Federation of Colored Women's Clubs was founded in 1906 to serve the needs of the state's African American women and children.

Each state had its own GFWC committee, and, along with her many other activities, Sara was the state chairman of the California Federation of Women's Clubs, also founded in 1890. The national organization took an interest in forestry, and women in two dozen states had devoted many hours to promote the value of native trees, even to the point of planting thousands of saplings.

By 1900 the Lemmons had published *Matchless Forest Endowment of the Pacific Slope, Particularly California* as a sixteen-page brochure, priced at twenty-five cents. This time they increased profits by not putting the book in stores and instead selling it directly from the Herbarium.

"You see," Sara told Mattie, "we are trying to be business-like for the future help on the big book [of conifers]—as well by these issuances to now meet the demand for information upon the great growing and awakened interest in forestry."

The brochure included an article specifically for women, "Some Hints upon Forestry" by Mrs. J. G. Lemmon. In it, she was far ahead of her time, urging female readers to educate themselves on the ecological benefits of responsible forestry practices: "Forestry teaches how to plant and cultivate trees, the suitable trees for reforesting and for ornamental purposes; the particular species of trees adapted to certain kinds of soil and elevations; it points out the enemies to tree growth, both animate and inanimate, and how to get rid of them."

"Why should woman be interested in the subject of forestry?" has been asked. Why should she not be interested in forestry since she enjoys the benefits with man and suffers with him in the loss of the trees?

Women can exert the most powerful influence to advance this noble work by becoming well-versed in the subject.

The couple had also published the fourth edition of the *Handbook of West-American Cone-Bearers*, which included five black-and-white reproductions of Sara's watercolors, as well as the full text of her article in the *San Diego Bee* about their trip to Baja to find Parry's Pinyon.

The book also solicited orders for two works "In Preparation": The first was for a five-hundred-page volume *West-American Forest Trees* with "100 or more characteristic full-page illustrations of the principal species, executed in the best style of modern art." The second call for orders was for Mrs. J. G. Lemmon's one-hundred-page *West-American Ferns and Where They Grow*, for one dollar. Neither book was completed.

But on November 25, 1900, JG inscribed a copy of the fourth-edition *Handbook* to "Mrs. Katharine Brandegee." She had been appointed co-curator of botany at the California Academy of Sciences in 1883 with Edward Greene—and had been the sole curator as of 1891 at the queenly sum of forty dollars per month.

This act of collegiality is the only evidence of interactions between the Lemmons and the Brandegees despite their shared passion for botany and common experiences. In fact, when Kate Curran married Townshend Brandegee in 1889, they too spent their honeymoon collecting plants— while walking five hundred miles from San Diego to San Francisco. But perhaps both couples were too busy to interact much. Or because both Kate and Sara had strong opinions and personalities—perhaps those got in the way of friendship? Kate once said Sara's "firm and aggressive attitudes had a powerful lifelong influence on her gentle, ungainly husband."

On the other hand, Kate herself was far from gentle. Greene, who was a minister in addition to being a botanist, once described her as a "she-devil." She was impatient and felt that Asa Gray took too long to publish descriptions and names of new species. So as acting editor of the *Bulletin of the California Academy of Sciences*, she provided a faster path to publication—which proved to be of benefit to the western botanists who chose to take advantage of it and a reason for disapproval for those who didn't.

In general, the fourth edition was well-received, and Sara wrote Mattie, "Great interest is being awakened in Forestry over here. In fact all over the country, an alarm is being felt for the future all through the wanton waste & destruction of our precious forest inheritance. We are constantly preaching the gospel of forest preservation."

Working on the *Handbook* had taken a tremendous toll on JG. At Christmas that year he told Clara Barton he'd been in bed for months and that Amabilis had been forced to do all the errands in town. Clara was still among Sara's closest friends even though they hadn't seen each other since the Webber Lake camping trip in 1886. In 1900 Clara was deeply embroiled in a Red Cross upheaval that would eventually lead to her resignation three years later—at age eighty-three.

In general, the tone of Sara's letters to Mattie, though always affectionate, is "proper" and somewhat stiff and reporter-like: In twelve hundred pages there's barely a single exclamation mark. But with "Sister" Clara, she relaxed and grew chatty, occasionally including an irreverent remark that revealed her merry side.

In one 1902 letter, while fiercely defending Clara's report to the administrators of the Red Cross, Sara allowed herself an epistolary giggle fit: "In short, it is competent, relevant and just hits every nail on the head that springs out of that *Board* of Control through its seasoning process. Ha! ha!! ha!!! ha!!!! ha!!!!! ha!!!!!! ha!!!!!!!"

Clara admired Sara's writing as well, which she described as "expressions of deep thought clothed in elegant and beautiful dress. I know that you are grateful many times for the power of direct expression which was given to you."

In 1901 Sara had yet another medical adventure. She'd been bothered by abdominal pain that she'd attributed to overwork—or perhaps as JG explained to Mattie, "She takes all pains and aches as concomitants of femininity, without protest."

Reluctantly, Sara saw a doctor, and JG told Clara, "Lo! the diagnosis was appendicitis—that frightful disease of a useless and always menacing organ." Because of the inflammation and the number of adhesions in the nearby intestines, Sara's June 3 emergency appendectomy lasted more than an hour. JG was relieved when she came out from under the chloroform at 3 p.m. and recognized his voice.

Everyone agreed how fortunate it was that she didn't have an acute attack while the couple was "in the depths of a forest or out on a wide desert of Arizona where death would result in a few hours."

While convalescing Sara worked on a different kind of project, a non-botanical book. At 466 pages, *A Record of the Red Cross Work on the Pacific Slope: Including California, Nevada, Oregon, Washington, and Idaho with their Auxiliaries; Also Reports from Nebraska, Tennessee and Far-Away Japan* was no small undertaking.

As JG told Clara, "despite the willing loss of her Appendix, as she says, she is able to turn off a vast amount of work in a day—and you should see the piles of letters, stacks of MS and proofs on tables, chairs, window seats and even invading one end of my botanical table!"

Sara was still an active member of the women's educational organization, the Ebell Literary and Scientific Society. On February 9, 1902, Sara told Mattie all about heading up an impending meeting, complete with decorations, music, light refreshments, a forestry lecture by JG, and one of her own watercolors of cone-bearing branches on display. She'd need to wear her best gown—one of a deep plum color with cream lace trimming and white kid gloves.

She was already thoroughly "fagged out" and added, "We often wish we were out in the wild woods. . . . There we can rest and let our souls loaf without social pressures."

In May 1902, California was the host state for the national biennial convention of the General Federation of Women's Clubs, during which hundreds of women would gather from all over the nation. Because she was the state chairman, Sara was in charge of the event and also had to make a formal address to the Los Angeles convention.

"Pray for me—" she entreated her always supportive husband. The event was complicated even more by a mix-up in delivery of the trunks belonging to many of the ladies "whose traveling dresses would never do for the reception," reported the local paper. "If these trunks are not all delivered by Thursday evening, there will be weeping and wailing and gnashing of teeth."

JG, reading the newspaper accounts at home in Oakland, wondered if Sara's trunk had arrived safely. Either way, he was confident "you did your work finely and in splendid style." He reminded Sara how proud he was to be her husband and assured her "you are becoming wiser and brighter every day of this cyclonic period. You have presented the strongest plea possible for the forest—and I fondly believe you have won the applause and sympathy of the mass of strong, influential women in that audience."

Then he urged her to have a good time "taking in all the sights and pleasures possible" before returning home:

> You know the character of the greeting which awaits you—my dear, loving wife: the heart and soul of my existence. . . .
>   Don't hurry—but be sure you will receive a hearty welcome by your tired, distracted—but proud and happy hubby.

In the summer of that same year, 1902, the Lemmons' next book, *How to Tell the Trees and Forest Endowment of Pacific Slope*, was published—and copyrighted in Sara's name. A slender volume of sixty-six pages describing cone-bearing trees, it included a section titled "Some Elements of Forestry with Suggestions by Mrs. Lemmon" on the vital importance of forest conservation, as well as black-and-white reproductions of her watercolors. In the preface she wrote, "It is to be hoped that this profusely illustrated little packet of leaves will be welcomed by all tree lovers and prove helpful in their becoming better acquainted with man's best friend on earth, the bounteous forest."

That book too was well-received, and although hardly unbiased, "Sister" Clara commented: "The country will not know until later the service that you and Lemmonia have been to it, but let us hope the grand old trees will know it and wave out their blessing, not only in the present, but a loving requieum [sic] long after you will sleep beneath their shade."

On August 11 Sara told Clara about their six-week "roughing it" trip the previous month to one of those bounteous forests, the King's River Cañon—with two hundred fellow members of the Sierra Club! She and JG had been dubious about traveling with such a large group, but those doubts soon became "a dim, misty memory. All went well—no sickness or accident during the month of high mountain climbing and exploring on & off dizzy trails, fording rivers, etc."

They slept under shelter only two nights, and she declared, "I boast of walking over 100 miles." After her appendectomy the previous year, her surgeon had advised her to trek on foot rather than ride horseback, so she "rode in all only about 4 miles and all the time gained in health and strength." JG rode most of the time—"you know he cannot walk much."

"The Sequoias—the noble Redwoods never looked grander in their Majestic Solitude," she reported and described their campsite: "We had a large space enclosed by ropes & hung about with pine cones, firs, etc. and our bed was spread upon pine and incense cedar boughs under the low sweeping fan-like branches of a beautiful silver fir facing a high mountain that looked down upon the swift flowing King's river. At night the distant stars would twinkle between the leafy branches & in the early morning the birds would sing their sweetest overhead."

At last, in December 1902, after four years of work, the Red Cross book was published, all 466 pages with its 238 photographs—including two of Sara herself.

On December 2 she told Clara:

The edition is limited to 1000, and there will be no future one as it was not electrotyped & the type is pied [meaning the type pieces were scattered]. It has been a tedious, long-drawn undertaking to gather and edit & arrange matter in all stages & conditions from so many directions.

You may be glad that I feel that the work is off my hands & seeming to give some, I may add, great satisfaction.

The exclamation is often "A beautiful book!"

That beautiful book came at a cost: Months later she confessed to Clara that because of the book deadline, "My eyes gave out through overtaxing them too soon after the attack of appendicitis & operation."

Fig. 48. Sara Lemmon at age sixty-five. She wrote her sister in December 1901, "I am having some pictures taken and hope to send you one early in the New Year to let you see how ugly I've grown." Photograph by F. H. Bushnell, from Mrs. J. G. Lemmon, Mrs. S. A. O'Neill, Mrs. G. S. Abbott, Mrs. L. L. Dunbar, Mrs. F. H. Gray, and Mr. J. G. Lemmon, *A Record of the Red Cross Work on the Pacific Slope: Including California, Nevada, Oregon, Washington, and Idaho with their Auxiliaries; Also Reports from Nebraska, Tennessee and Far-Away Japan* (Oakland: Pacific Press Publishing Co., 1902).

Her doctor ordered her out of doors, so Sara took up gardening as therapy—in typical Sara style: She planted fifteen different vegetables, along with nine varieties of flowers, including, she told Clara, a "rich mass of scarlet geraniums that could be seen a mile away from the summit of Piedmont hills." She added, "The *Tagetes Lemmoni* [*sic*], shrub marigold, is coming into full bloom. When seeds are ripe, I will send you some if you so desire. It is native to Arizona, discovered in one of our trips there." According to one source, the progeny of that pungent garden wildflower was introduced to the nursery trade and traveled as far as England by the early 1900s.

The following spring, March 3, 1903, the *Oakland Enquirer* published a special dispatch with a front-page headline:

THE GOLDEN POPPY IS OUR EMBLEM

Pretty and Impressive Ceremony Follows the Reading
of the Message
Mrs. J. G. Lemmon of Oakland Is Complimented
on Her Work in Relation to the Measure

On this third attempt, Mr. F. Smith introduced the bill he'd written with J. A. Bliss into the assembly. The two men had even arranged to have a poppy placed on the desk of every legislator—but owing to a train wreck, the flowers didn't arrive on time.

By now George C. Pardee, a Republican doctor and former Oakland mayor, had been elected governor of California. He was also a supporter of conservation issues and was no doubt familiar with the work the Lemmons had done. In recognition of being "the one largely responsible" for ten long years, Sara was awarded the "magnificent" Bald Eagle quill-feather pen used by Governor Pardee to sign the bill making the golden poppy the state flower of California.

Of course the momentous event required a speech from Sara. She began, "I cannot be expected in the time allowed to attempt any lofty eagle-flights of rhetoric. I will only give a brief account of this important legislation." She then recounted the struggle, lauding the ladies of all the districts as well as Senator Smith and Assemblyman Bliss.

She concluded by saying, "I shall ever cherish this sacred souvenir."

The *Enquirer* reported Sara Allen Plummer Lemmon then "pinned on the lapel of Mr. Bliss's coat a badge displaying a cluster of poppies done in oil color—the work of her own hands."

22

"Safe—Tho' Tremendously Shaken"

Arizona and Oakland, 1903–6

→ WITHIN HOURS OF THE golden poppy legislation, a triumphant Sara wrote Mattie: "You should see the gold-mounted engraved souvenir made from the quill of the American Eagle by which Governor Pardee signed the bill legalizing our State Floral Emblem."

She belonged to many civic, political, and environmental organizations, and for months it seemed every one of them hosted a reception in her honor.

As if the Lemmons' lives weren't already jam-packed, Sara's niece Mattie St. John and her husband, Charles, were about to visit. Sara lamented that the house wouldn't be "spick and span." Her nephew Charley was still living with them at the Herbarium and planned to take time off to spend with his sister, saying, "Only think of it, I haven't seen Mattie in over four years!"

Recently, he'd given Sara and JG a fright when he'd had "quite a sick attack" that they feared was pneumonia. Sara went to work making him perspire using six applications of a rum sweat—which had the desired effect, even though "I nearly burned the bottom out of a cane seat chair—everything just *smoked*," she told his mother. Fortunately he recovered before Mattie and Charles arrived, and afterward everyone agreed the sole defect of the visit was that it was too short.

By 1903 San Francisco was the ninth-largest city in the United States and exploding as a trade, cultural, and financial center. Even Oakland, viewed through the Herbarium's many windows, was buzzing. Sara wrote: "We are busy inside our plain house and full of confusion & bustle without: Street grading and paving is going on just a block above us, three new two story houses are building, and things out of doors about us looking promising," and adding, "a great time ahead with the President's visit."

The Bay Area was indeed bustling with preparations to salute President Theodore Roosevelt in San Francisco on May 13, 1903. That morning his carriage rolled majestically down Van Ness Avenue, escorted by the all-Black Ninth U.S. Cavalry Regiment as three thousand school children waved flags—and for the first time, a president was captured in a motion picture. From there he moved to the Presidio Military Reservation, where he reviewed the troops, under the command of Major General Arthur MacArthur (father of Douglas MacArthur).

Roosevelt took luncheon on the verandah of the Cliff House, feasting on salted almonds, Toke Point oysters, and even a second helping of the "Filet of Sole, a la Cliff House," followed by strawberries, ice cream, and "Fancy Cakes." The menu was of course richly adorned with illustrations of the golden poppy, California's brand-new floral emblem.

Spring brought disastrous flooding to Mississippi that required much help from the Red Cross. Clara Barton oversaw the rescue efforts, and both Sara and JG offered her their organizational skills. Fortunately, they weren't needed.

Despite their years—Sara was now sixty-nine, JG seventy-three—and fragile health, they weren't about to retire. They envisioned their biggest book yet, their legacy work that would be an even more complete volume covering all of western botany.

So in June 1905 they left Oakland ready to gather plants, photographs, and sketches with the ambitious intention to visit every Pacific Slope forestry district, including California, Arizona, Texas, New Mexico, the Mexican republic, Nevada, Oregon, Washington, British Columbia, and Alaska.

They began the journey with a stop at Santa Barbara, where, on June 8, the local paper reported, "Mrs. Lemmon is herself a botanist of unusual ability, but she will be best remembered as the woman who worked for ten years to secure the official adoption of the California poppy as the state flower." The reporter also remembered her as "one of the prominent factors in local society in the palmy days of the Santa Barbara college and at the time of the formation of the educational standards of the town."

By Friday, June 30, they'd moved on to Tucson, where they stayed at the Willard Hotel. One can only imagine the chuckles of the now long-married couple as they read the gushing *Tucson Citizen* article:

> One morning twenty-five years ago just as Old Sol was climbing over the Rincon mountains, J.G. Lemmon a young botanist of Oakland, Cal., arrived in Tucson. With him was his blushing bride of a few days. Life seemed to the young couple one continuous song and the sunrise seemed the most beautiful they had ever seen.
>
> This morning J.G. Lemmon of Oakland arrived with his bride of twenty-five years, just as the sun was peeping forth. The sunrise was fully as beautiful for them as it was a quarter of a century ago.

The Lemmons had chosen to mark their silver wedding anniversary in Tucson and had "hunted up" their old friend Emerson Stratton, the rancher who'd led them up the Santa Catalinas on their 1881 honeymoon. In the intervening years, George Roskruge, the uncle of Stratton's son-in-law, had become the Pima County surveyor. In 1904 he made a new county map and added a name for the highest point in the Catalinas.

To JG's chagrin, Roskruge first labeled the prominent feature as "Mount Lemon."

Now the aging but still sturdy trio hiked and rode to the peak that had been (correctly) named Mount Lemmon. Years later Emerson reminisced to his daughter about the two trips. In 1881, "from Carter's camp we went to the highest peak of the Santa Catalina Mountains and christened it Mount Lemmon in honor of Mrs. Lemmon who was the first white woman up there." Then he added, "Just twenty-five years later on their silver wedding anniversary, Dr. Lemmon and his wife returned to Tucson, hunted me up, and we again climbed the mountain."

Typically modest, Sara neglected to mention the mountain's namesake in her July 25 postcard to Mattie: "Both well & happy on this our Silver Wedding Journey. Mr. E. O. Stratton who accompanied us on this North

Fig. 49. Sara's postcard, dated July 25, 1905, to Mattie describing their silver wedding anniversary trip up Mount Lemmon, the peak named for her. Photo by author. Original at the UC and Jepson Herbaria Archives, University of California, Berkeley.

Santa Catalina trip & to Mt. Lemmon is with us here again. We sleep in the open air beside a delicious mountain stream that sings our nightly lullaby. Five burros take us over the rugged mountain trails."

From Tucson, the Lemmons revisited Fort Bowie, the Huachuca Mountains, and then El Paso before traveling deep into Mexico. In August Sara sent a postcard to Mattie from Guadalajara, in the state of Jalisco, mentioning JG's "inflammation of the colon" and hoping they'd be able to continue on to Mexico City.

That same week JG wrote Emerson Stratton (omitting any reference to his colon) about the country's botanical delights—"rare trees and new plants on every hand"—and especially about the hospitality of no fewer than eight Mexican governors who'd provided them with fine riding horses for their work—and even a sailboat! They were off to Mexico City, but he assured Stratton they wouldn't "stay long in the town, our business is with the forests."

Although no other letters from that trip have survived, a botanical remnant has. One specimen of lip fern, *Cheilanthes viscosa*, still remains at Chicago's Field Museum. According to its identification label, the fern was collected "in the vicinity of La Venta, State of Jalisco, Mr. and Mrs. JG Lemmon, 1905."

In 2012 plant taxonomists examined the entire *Cheilanthes* genus using molecular genetics. They moved *Cheilanthes viscosa* into its own new genus. The plant's scientific name is now *Gaga kaulfussii*. Georg Friedrich Kaulfuss was a noted German botanist who specialized in ferns and died at age forty-four in 1830 after naming some two hundred plants.

The plant's common name is Lady Gaga Kaulfuss lip fern. The taxonomists explain: "The genus Gaga is named in honor of the American popsinger-songwriter-performer Lady Gaga, for her articulate and fervent defense of equality and individual expression in today's society. Because Lady Gaga speaks to the need for humanity to celebrate broad differences within its own species, we hereby provide her with a scientific namesake that characterizes the struggle to understand the intricate biology underlying cryptic patterns of diversity."

The authors also point out that the letters GAGA are a genetic sequence pattern that appears only in the plants assigned to this new genus.

The couple returned home to Oakland, where, within the next year JG's health faltered even more. Sara shouldered more of his research, leading some experts to suspect she'd done much of the work attributed to him all along.

In 1906, LeRoy Abrams, a botanist and professor at Stanford University, encountered a specimen the Lemmons had collected more than twenty years earlier in Mohave County, Arizona, near Kingman. He named it *Penstemon plummerae* in Sara's honor. Sadly, the species, commonly known as yellow bush snapdragon, was later folded into the *Keckiella* genus and lost its nomenclature connection to Sara Plummer.

In that same year JG's text and one of Sara's illustrations were published in the March issue of *Out West: A Magazine of the Old Pacific and the New* along with a full-page photograph of the two of them in the Lemmon Herbarium.

On April 16, 1906, they gave a lively, well-received presentation on their Mexican travels to the Oakland chapter of the Women's Alliance.

The next evening didn't seem unusual. Enrico Caruso, of the Metropolitan Opera Company, gave a "towering" performance in the lead role of Don José in *Carmen* at the Grand Opera House in San Francisco. A three-alarm fire broke out on Market Street. The wind shifted directions and blew onto the city from the sea.

At midnight James Hooper, a journalist with the *San Francisco Call* noticed the horses at the local livery stable were unusually restless and wrote, "I heard a score of hoofs crashing in tattoo against the stalls." During the wee hours in the Mission District pound, the dogs barked loudly enough they awakened a priest, Father Charles Ramm. At 5 a.m. a milkman's horse was so anxious the man could hardly get the harness adjusted.

Then, at exactly 5:12 a.m., April 18, 1906, the San Andreas fault convulsed, and the San Francisco Earthquake slammed much of the Pacific Slope.

Terra firma billowed into roiling waves of earth two and three feet high. "The motion seemed to be vertical with a rotary twist, like a French cook tossing fish in a frying pan," wrote James Hooper, the *Call* reporter.

Whole buildings danced—then collapsed in a roar, accompanied by the jangling of every church bell in town. People still in their pajamas, some with faces smeared with shaving cream, poured out into the streets.

Tremors from the main shock lasted forty-two seconds and were felt from Eureka, just south of the Oregon border, to the Salinas Valley, a total distance of three hundred miles. Sixty-five miles north of San Francisco, the shaking flattened Santa Rosa's downtown.

In 2015, Ruth Newman, the oldest remaining survivor of the earthquake, died at age 113. In her *New York Times* obituary, Newman's daughter said her mother's memories of the earthquake still remained vivid a century later. She lived seventy miles north of the city on a ranch near Healdsburg, California, and remembered her "grandmother being upset because they had just milked the cow earlier, and she had separated the cream and all and put it in containers that got thrown to the floor."

The Richter magnitude scale wouldn't be developed for another thirty years, but contemporary estimates put the measurement at 7.9 to 8.3.

The earthquake itself was trauma enough, but within seventeen minutes, residents reported nearly fifty fires—just in downtown. Many of the city's gas mains ruptured, bursting into firestorms. All the water mains shattered, leaving the city's fire department helpless.

The inferno raged for four days, destroying more than 80 percent of the city, at a cost of $11 billion in today's dollars.

Because no one considered the residents of Chinatown worth counting, the actual number of deaths is unknown. But estimates are close to three thousand.

Of San Francisco's population of 410,000, somewhere between 227,000 and 300,000 people were left homeless. Around half of those fled to Oakland and Berkeley—and some ended up at the Lemmon Herbarium.

"Thank Heaven we are safe," Sara scrawled hastily on an April 24 postcard to Mattie: "Severe shocks & damage, but we do not think of selves with stricken, ruined San Francisco. We are housing 9 refugees, &

more to come. How our hearts ache for every one there in S.F. As far as we know, all whom you & we know escaped. Some lost everything. The papers *cannot* exaggerate—however much they may mistake facts. It is terribly horrible. We are busy every minute caring for the distressed ones."

Estimating the fire damage was made more complicated by local insurers who had written policies for damage by fire—but not for earthquakes. Consequently, some residents set fire to their own houses for the insurance money.

The Lemmon Herbarium had its own share of damage. On May 1 and 4 Sara snatched a few moments to write Mattie another postcard. The inspector had stopped by and told the couple that the chimney had to come down. They'd also have to replace all the cracked window glass and repair all the walls and ceilings.

However, the fifty-dollar cost "seems trivial compared to the personal, helpless distress of the thousands with no roof to shelter & no clothes to change. We have kept open house for nearly a dozen refugees who continuously lodge and eat with us & others come and go for chatter. We cannot go on with our regular work or duties."

She added, "JG is ill with pleurisy and cold." They were both ill with heartbreak.

In addition to burning twenty-five thousand buildings over 490 city blocks, the fire gutted the California Academy of Sciences, by now the largest botanical collection in the western United States. It was also where Sara was the first woman allowed to speak back in 1881, where she'd met Sir John Hooker and Dr. Asa Gray, and where so many of her paintings and the couple's field notes, journals, and specimens were housed.

In 1894 the remarkable self-taught Canadian-born botanist Alice Eastwood had replaced Kate Brandegee as curator of botany at the Academy. Even as the quake still shook the streets that terrifying morning of April 18, Eastwood hired a passing wagon and with an assistant's help, she dashed in and out of the building, rescuing as many rare and type specimens as she could. All the while, her most treasured possession, a Zeiss microscope lens, lay nestled safely in her pocket.

Later Eastwood described the spirit of San Francisco as the tremors subsided: "Nobody seemed to be complaining or sorrowful. The sound of trunks being dragged along I can never forget. This seemed to be the only groan the city made."

The fire eventually claimed both the Academy and Eastwood's own house. Yet she wrote, "The kindness of my friends has been great. . . . I feel how very fortunate I am; not at all like an unfortunate who lost all her personal possessions and home."

Among those friends was JG, who sent a postcard on May 4:

Dear friend,

How and where are you? We've expected to hear from you any day. Do of course know that the great herbarium—the work of your hands—was destroyed, but we hope you saved personal articles, unpublished notes, etc. Both send warmest sympathy.

Yours always,
J.G. Lemmon

The botanical world was fortunate: Alice Eastwood had managed to save 1,497 rare and type specimens. However, tragically, not one of Sara's paintings or JG's specimens and photographs were among them. Although Sara never complained in her letters, both she and JG had to have been shattered by the loss.

23

"I Feel So Helpless and Alone"

Oakland, 1906–12

⇥ FOR MONTHS FOLLOWING THE earthquake, Sara and JG set botany aside and continued to provide lodging, meals, a gathering place, and whatever other comforts they could for displaced victims. In September, four months after the quake, they even applied for a permit to add a cottage to their backyard.

By July 1907, life was approaching whatever "normal" would become for the resilient and adaptable Bay Area residents. Sara told Mattie: "We have had so many interruptions through the earthquake and fire, we lost over a year of time." Even their inventory of published books was reduced as they'd had "quite a stack of books" for sale in a San Francisco store that had been destroyed.

Undaunted, JG bent over his field notes, and Sara buried herself in watercolors to illustrate the next book. "We are as busy as bees and happy in the work."

Like the Lemmons, everyone in the surrounding urban areas was united in resolve to rebuild their lives. "Everything in the way of business is flourishing despite graft, strikes, etc.," said Sara. "The spirit of progress is in the air & will not be held back. It is a wonder how Oakland & San Francisco are growing."

Sara and JG had repaired the Herbarium damage, and the grounds now looked so lovely that tourists stopped to gaze at the 150-foot-long golden mass of California poppies. Sara told Mattie, "We feel proud of that bed of Eschscholtzia."

Throughout her life, particularly when they were home, Sara's Unitarian church life had been a mainstay. She'd told Mattie about a new minister,

Reverend William Day Simonds of Seattle, who'd arrived in Oakland: "We hope great things from him in calling together the luke-warm but liberal people who should fill our beautiful church to overflowing in such a large community. Somehow Unitarians are not aggressive and proselytizing, like most others."

The strategy must have worked; a year later, Sara would report, "Church full. Reverend Simonds brilliant as ever." According to its website, the First Unitarian Church of Oakland is still at the corner of Castro Street and Fourteenth Street and continues to be "a beacon of liberal religious acceptance, tolerance and social activism."

Ever the faithful aunt and in loco parentis, Sara also reported to Mattie that "your boy Charley looks fine, seems happy & is prospering." Sara was thrilled that her nephew was involved in real estate: He'd bought several lots and added a bungalow to one of them. Even though he now had his own place, he attended church with his "Aunt Sadie" and stopped by frequently to check on them both and to share any family news from Dover.

Sara and JG would soon be relying on him more and more. In the meantime, of course Sara couldn't resist reminding Mattie how much all three of them were counting on her coming to Oakland for the winter.

That same month, July 1907, the *San Francisco Call* announced the Lemmons' latest ambitious project, their "big book" that would describe in detail all the timber from Alaska to the Mexican line.

The reporter went on to describe the pair as "intrepid travelers" who for weeks "slept in the open air of the desert or among the giant forests of middle Mexico. The barren mountains in Arizona and New Mexico held no fear for the aged botanists. Professor Lemmon at that time was past his seventieth year while Mrs. Lemmon was aged more than 60 years."

That same year Mayor Frank Mott (known later as "the man who built Oakland") appointed Sara and JG as delegates to represent the city at the Fifteenth National Irrigation and Forestry Congress in Sacramento. Although they were in "declining health," the topic was too important to them to miss.

Perhaps it was that awareness of declining health that spurred the couple to make a trip to Oakland's Mountain View Cemetery one day that fall. There they rambled through the manicured grounds, eventually selecting and paying for a burial plot. Designed by Frederick Law Olmsted, the same landscape architect who designed New York City's Central Park, the graveyard would be the final resting place for many who'd been a part of the Lemmons' lives: Charles Crocker, who'd made their Southwestern explorations possible by giving them railroad passes; Governor Pardee, who'd signed Sara's poppy bill; Sara's friend and doctor, Chloe Annette Buckel; and their good friend Ina Coolbrith, California's first poet laureate, to name just a few.

The following year, in 1908, JG published "Notes by a Pioneer Botanist" in the April 14 edition of *Muhlenbergia*, a monthly botany publication. In it he reminisced about his beginnings as a botanist, recuperating from the horrors of Andersonville at his brother's home in Sierraville—and his intense joy when Dr. Asa Gray named six plants in his honor.

It would be his last publication. He told Mattie a month later that he was still working on the "interminable mass" of their manuscript and "really the end seems in sight." But "often I am discouraged—principally on account of failing eyesight. My left eye is weak and wanders more and more from its mate, which the right one though retaining its strength, yet becomes weary with doing all the work, and so is in pain much of the time, and at night the aching allows but little sleep."

An addendum from him shows that at long last, Mattie succumbed to Sara's badgering and did indeed come for a "few weeks visit in which to recall the blissful past and plan for the (perhaps) bright future." It seems she even invested in one of Charley's houses—Sara must have been so pleased.

In early August Sara and JG reluctantly concluded his eye strain was too severe for them to travel. Instead they resolved to focus on the book with the goal of publishing it in the spring.

On August 31, Sara began a postcard to Mattie saying that JG's eyes had improved, thanks to twice-weekly "electric treatments" and restricting his work to half-hour sessions. But now he was under treatment for a constant pain in his lungs and a hacking cough.

Even more reluctantly, they'd accepted that they couldn't push any harder on the book.

As often happened, Sara got distracted while writing that postcard, but she returned to it September 8 with more sad news: Their dear friend and Sara's painting mentor, Adelia Gates, had gone blind and now required a full-time nurse. Sara told Mattie that her longtime friend "seems to be fading slowly away. She can never again see the light of day but is utterly resigned to her fate."

As if managing the Herbarium and JG's health wasn't enough, Sara went to Adelia's every week to "attend to her affairs as her right hand."

By late fall, life had settled into a routine. Oakland's usual dry summer weather had broken, and November 22 was a cloudy Sunday. "Take Me Out to the Ball Game" was a new top song in the country, and strains could be heard on pianos and phonographs everywhere in town.

Charley stopped by to see Sara and JG that morning and, while there, grabbed some Herbarium stationery to write Mattie a quick update on their health. Sara had decided to skip church because JG had been in pain all week and was spitting up blood.

On the previous Thursday night his already-frail condition had worsened. Dr. Frank Adams, the county's best doctor, according to Sara, was treating him for congestion of the lungs and "valvular heart trouble." But that Thursday, just as JG became even sicker, Dr. Adams unexpectedly died.

Sara asked a Dr. Kelley, who lived across the street, to come examine her husband. She also hired a trained nurse to stay at his bedside during the day. She was exhausted: Each night she had watched over him from 10 p.m. until 5 a.m.

"Uncle John is not out of danger yet," Charley wrote his mother that day, "but he is gaining. He had a good nap this afternoon." Even so, he said, Sara had telegraphed JG's older brother who still lived in Michigan that he should come although the two men hadn't seen each other in years.

Charley added, "Aunt Sadie has not been very well lately" and that she was tired and worn out from staying up three consecutive nights with JG. Two days later, on Tuesday, November 24, Sara wrote Mattie:

Just before mailing—I watched over JG every *moment* from 11 p.m. to 8 a.m. to let the nurse have more sleep. The hours were anxious ones. He is *very* weak, and I have but a shadow of hope that he can recover. He has two things to contend against—valvular heart trouble & pneumonia & is too weak to lift his head or help himself. He is fed from a teaspoon & given a little milk, broth, white of egg, rice, water, etc. no solid food, etc. But for his being so abstemious, the Doctor says he does hope that he will pull through—

Will write again—
Sara

Hope wasn't enough. Within a few hours, pneumonia stole Lemmonia from his loving Amabilis two days before their twenty-eighth wedding anniversary.

The "shock of sorrow" slammed into Sara. Yet she still had to organize all the funeral and burial arrangements, comfort their friends, and try to maintain a gracious front. Not surprisingly, in December she collapsed from exhaustion, grief, "a heavy cold, acute bronchitis, and a general breakdown."

On December 14 she finally dragged herself out of bed, dressed, and wrote her brother Osgood to thank him for managing some pension details. She apologized for "burdening" him with the task and went on to describe the funeral, held Saturday, November 28. Crowds of friends and well-wishers packed the First Unitarian Church for the service. The *Oakland Tribune* reported the ceremony was "martial in nature" as JG wished and under the auspices of the GAR's Admiral Porter Post, where he'd served as chaplain for eight years. Eulogies included a "glowing tribute" from Sara's lifelong friend Professor Albin Putzker (the University of California's first official professor of the German language), as well as readings by the Congregationalist minister Dr. J. K. McLean and the much-admired Reverend Simonds. Botanists Professor Willis L. Jepson and T. S. Brandegee were among the pallbearers.

At Mountain View Cemetery, the post commander had lined the open grave with

> pearl-white cloth and that festooned with smilax and ferns and La France roses—not a particle of earth in sight. There were quantities of beautiful floral emblems, and one member of the Red Cross had made a *red cross* on a white ground, all flowers, about 2 feet square. Dr. Jepson of the State University sent a large wreath of pine cones & green leaves—very beautiful—another an olive wreath, a laurel wreath.
>
> Everything was done by the public to show their love, honor & respect.

For days after the funeral, tributes, honors, and kindnesses flooded into the Herbarium. One woman, who knew Sara before she'd married JG, happened to be visiting Sonoma County and heard about his death over Thanksgiving dinner. Instantly she left the table, took the next train to Oakland, "and says she will stay with me till I get better—a *real* friend in need."

More than one hundred letters "expressive of sympathy and great appreciation of my comrade-husband's world's work" poured in. The mayor sent condolences, as did Luther Burbank, the horticulturalist now best remembered for the Burbank Russet potatoes, the variety used in McDonald's french fries.

John Gill Lemmon was a complicated mix of brilliance, determination, and eccentricities, all made more extreme by wartime trauma. Charles Lummis, former editor of *Out West* magazine, wrote, "Professor Lemmon was one of the makers of history and scholarship in this state."

One of the most comforting condolences to Sara was their friend John Muir's sweet, heartfelt note of condolence:

Dear Mrs. Lemmon:

Only illness prevents me from being with you in your bereavement this sad lonely day. Last goodbyes are always sad. Yet the memories of your walks with your loving husband, wandering free in God's woods through so many long happy years must be a blessed & enduring consolation.

I pray Heaven bless & comfort you & am with sincere sympathy faithfully yours,
John Muir

In her letter to Osgood, Sara despaired about paying for the doctors, undertakers, grave opening ("How grim!"), and more: "Now I'm in a pinch that could I have foreseen would make me wince more than I do now—when I feel so helpless and alone!" But, ever practical, she'd already put the Santa Barbara lot she'd held for thirty years on the market and made plans to rent out the Herbarium's extra rooms.

How she wished Osgood could have gotten to know JG: "modest and retiring, but ever on the right track, never pushing himself ahead—you would have loved him for his high principles, clean life, bright intellect & sterling worth."

In his reply, Osgood mused that in most couples someone has to die first and suggested that, in their case, perhaps it was just as well it was JG.

Sara responded: "In thinking it all over, I do agree with you that I can be better left to bear the ills of life than could JG. He was so frail I have often found myself in the past, transfixed, looking into his thin worn face and wondering how he could ever have the courage to live and do anything—broken as he was through the war and dreadful Anderson [sic] prison-pen experiences."

She concluded, "I am glad that I can look back and half realize that I've had the privilege of walking and working beside such a nature-nobleman for 28 years. For that I'm comforted, but for that I feel the loss inexpressibly."

24

"Partners in Botany"

Oakland, 1908–23

→ AFTER JG'S DEATH IN November 1908, Sara struggled to carry on a life without him. She continued living in the Herbarium, sharing the space with the ghosts of their twenty-eight years together, company that was a mix of mournful and reassuring. She dropped her numerous memberships, living a quiet, secluded life. Charley continued to look in on her most evenings.

At last money was less of a worry, as her insistence on buying real estate paid off: In 1909 she sold the Santa Barbara lot for $300—about $8,100 in 2020 currency. She also received $12 per month as JG's pension from the government.

In 1910 friends persuaded her to revisit her old haunts in Santa Barbara after an absence of thirty years. The trip energized her, and on March 10 the *Weekly Press* published her lengthy article "Santa Barbara's First Library Efforts and Other Historical Sketches."

In it Sara wrote: "I am urged by former friends and newfound ones to give a personal and reminiscent sketch of why and how I came to Santa Barbara, my first impressions of the place, whom I met, how I regained my health and what I did to establish Santa Barbara's first public library. In an unguarded moment I consented, not realizing the difficulty of stirring the memory to a sufficient degree for the requirements of such an article. This is my foreword of apology for some inaccurate dates that are bound to occur."

Her article elaborated on many details of early Santa Barbara, explaining how dozens of streets earned their names. She even included an alphabetized list of more than one hundred early library donors and patrons.

Later that year, in the summer of 1910, she traveled to Europe with fifteen friends and family members, including her sister Mattie with her three now-grown children, Mattie (Everett) St. John, Charley Everett, and Sadie (Everett) Humphries, and Mattie's three grandchildren, Harold St. John, Everett St. John, and Vivian St. John.

The two-month trip was part of a Unitarian delegation to Berlin for the Fifth International Congress of Free Christianity and Religious Progress. The group set off from Canada, and their tour included stops in England, Ireland, Holland, Austria, Hungary, Italy, and elsewhere in Germany.

Sara was seventy-four at the time—and apparently a force to contend with. The family remembers her as being "irascible" at Kew Gardens, and one note in her grandnephew Harold's journal mentions spending most of an evening "trying to straighten out Aunt Sarah." Between his parents' deep interest in natural history and the influence of Sara and JG, young Harold's fate in the life sciences was happily sealed in childhood. By this time, he was a student at Harvard University studying botany under Asa Gray's successors.

Months after the trip, on January 28, 1911, Sara wrote him a postcard, saying, "What a great time you're having botanizing!" Still doing a little botanical work herself, in 1912 she contributed a list of ferns in Yosemite to help the Sierra Club.

Sara was such a habitual correspondent it seems likely that she continued to write letters after JG's death, yet few have survived. Perhaps she felt Charley was relaying everything that needed to be said to Mattie—or perhaps the two sisters actually spoke by phone? Oaklanders had been enjoying a rudimentary telephone system since 1881. By 1911 Oakland residents and businesses numbered 22,085 subscribers. It wouldn't be surprising if she were among them.

One 1912 letter from Sara has survived, to a Mrs. Hastings, in which she defended her own legacy and what she'd learned since JG's death:

I've been driven to study the *human motive* since I've stood alone during the four past years. I've learned much in meeting all sorts of obstacles & by stern will power to dare & do when I've become sure I'm in the right.

I've overcome all obstacles thus far & by all the talent—much or less given me to develop, I mean to keep on to the end, ever trying to keep the inward light burning. . . .

I mean now to defend myself in justice to myself & friends for protection that my influence & work may not be destroyed.

The following year those around her acknowledged that Sara's "inward light" was dimming. In fact, Mattie believed Sara's downward spiral actually began with the shock of the 1906 earthquake.

Then, as now, dementia was a harrowing and debilitating disease. By 1915 Sara's neighbors considered her "exceedingly eccentric." In October the *Oakland Tribune* reported that she'd blocked the passage of six streetcars as she insisted "upon sweeping the street between the car tracks with a large broom."

Charley was no longer able to handle his aunt alone, and the long-suffering Mattie traveled to Oakland to help. Although they tried keeping her at home, Sara became more than irascible—she was violent, irrational, and abusive to her family members. Her condition deteriorated still more until this brilliant, distinguished, dignified, elegant woman became "noisy, restless, dangerous." She refused to take any medicine, sure that people were trying to poison her. The lifelong Unitarian churchgoer became "vulgar and profane" and had "to be watched to prevent her from leaving her home unclad."

Finally, at 7:30 p.m., April 22, 1916, Charley Everett signed a form committing Sara to Stockton State Hospital for "senile dementia." The commitment paper identifies her as "Housewife & Botanist." She was seventy-nine years old.

Ironically, Stockton State Hospital was the same facility where Sara had escorted her sister-in-law, Rebecca Lemmon Olesen in 1899. In the 1880s the conditions had been deplorable: overcrowded with so little heat that patients shivered as they ate. By the time Sara arrived in 1916, however, "insane" patients were treated more humanely—even though

Fig. 50. The gravestone in Oakland's Mountain View Cemetery that reads "Partners in Botany." Photo courtesy of Amy St. John, great-great-grandniece of Sara Plummer Lemmon.

at the time California led the nation in promoting sterilization of the mentally ill. A new building had been constructed at Stockton in 1908 with separate male and female units and a dayroom for each. There was even a library for the female residents, donated by Miss Dorothea Dix, the famous crusader for the mentally ill.

Visitors reported that Sara "remained in fairly good health, spending a good deal of her time reading."

After six years in the Stockton hospital, Sara Allen Plummer Lemmon died at 6:30 p.m., January 15, 1923. The cause was listed as general arteriosclerosis. She was eighty-six.

Her funeral took place two days later. Sara's obituary in the *Pacific Unitarian* stated that her death "recalls to memory two savants and public-spirited citizens who rendered valuable services to natural sciences and the higher educational and social interests of Oakland and the State of California." She was buried in Oakland's Mountain View Cemetery, next to her beloved Lemmonia.

The year 2003 was the centennial of Sara's success in making the golden poppy California's state flower. In commemoration of that event, the Mountain View Cemetery's docent group helped facilitate a new marker for Sara and JG. The cost of the gravestone was donated by Omega Monuments.

On it are engraved the words "Partners in Botany."

The actual balance of that partnership between JG and the woman known to the world as "& wife" is still—and probably always will be—up for question. But just as Sara and JG had stood on the shoulders of the botanical giants before them like Asa Gray, Cyrus Pringle, and Charles Parry, so would future ecologists, like the great Forrest Shreve twenty years later, rely on the work of this determined couple.

Yet, as fellow botanist Alice Kibbe once wrote, "Sara was a brilliant and charming lady. . . . We will perhaps never know how much of the Lemmons' joint work was her doing, but we might suspect that it was considerable."

Epilogue

→ WHAT LITTLE IS LEFT of Sara Allen Plummer Lemmon's art lives on at the University of California, Berkeley, just a few blocks from where the Lemmon Herbarium used to be.

Sara's grandnephew, young Harold, went on to excel in botany and eventually became a professor at the University of Hawaii. There, the Harold St. John Plant Sciences building is now named for him. On his death in 1991, Harold left behind two boxes of Sara's paintings from the Lemmons' 1884 trip to Arizona.

In 2017 Amy St. John, Harold's granddaughter and Sara's great-great-grandniece, donated those two boxes of watercolors to the University of California and Jepson Herbaria Archives. The paintings, some of which are reproduced in this book, are still exquisite more than a century later, despite the effects of humidity and sometimes massive insect damage. They are much too fragile to ever be put on display or to travel. But conservation efforts are underway to forestall any further damage. Partial proceeds of this book go toward that conservation effort.

Acknowledgments

⇥ IT'S TAKEN A CONSIDERABLE village to build this book. I may be the mayor, but I couldn't have done it without the help of many other citizens.

First among them is my loving husband, Dave Peterson, who's so much more than the "go-fer-chauffeur" he claims to be. We joke that he married two women: Without his unwavering support and enthusiasm, Sara's story and I would still be trapped in a very long Scrivener binder of references.

Biographers always recite their gratitude to archivists, curators, and librarians—and I'm right there with them. I'm especially grateful to archivist Amy Kasameyer who made it possible for me to spend many hours in the University of California and Jepson Herbaria Archives at the University of California in Berkeley.

Other archivists, librarians, and curators who have my appreciation include Rachael Black, librarian and archivist, Library and Archives, Arizona Historical Society, Tucson, Arizona; Lisa DeCesare, head of archives and public services, Botany Libraries, Harvard University Herbaria; Chris S. Ervin, archivist, Santa Barbara Historical Society, Santa Barbara, California; Veronica Furlong, project archivist, National Park Service Regional Office, Western Archeological and Conservation Center, Tucson; Stephen Gregory, museum technician, Fort Huachuca Museum, Sierra Vista, Arizona; Nancy L. Janda, assistant archivist, Hunt Institute for Botanical Documentation, Carnegie Mellon University, Pittsburgh, Pennsylvania; Helen Jentzen, reference librarian, Mills College, Oakland, California; Rebekah Kim, librarian and archivist, California Academy of

Sciences, San Francisco; Thomas Labedz, collections manager, Nebraska State Museum, Lincoln, Nebraska; Michael Maire Lange, copyright and information policy specialist, Office of Scholarly Communication Services, University of California, Berkeley; Rebecca Leung, archives and manuscripts librarian, F. W. Olin Library, Mills College, Oakland; Nora Lockshin, senior curator, Smithsonian Libraries and Archives, Washington DC; Louise Maxwell, Llewellyn Park Historical Society, Orange, New Jersey; Lisa Prince, reference archivist, California State Archives, Sacramento; Perri Pyle, archivist and librarian, Arizona Historical Society, Tucson; Rebecca Smith, head of reader services and technical processing, the Historic New Orleans Collection, Williams Research Center, New Orleans, Louisiana; Debra Trock, director of science collections, California Academy of Sciences, San Francisco; Marya Van't Hul, curator, Natick (Massachusetts) Historical Society; Bill Westheimer, webmaster, Llewellyn Park Historical Archives, West Orange, New Jersey; Amy Weiss, collections manager, William and Lynda Steere Herbarium, New York Botanical Garden, the Bronx, New York; Molly Wetta, library services manager, City of Santa Barbara Library; J. Dustin Williams, digital assets administrator, archivist, and senior research scientist, Hunt Institute for Botanical Documentation, Carnegie Mellon University, Pittsburgh, Pennsylvania; and Leon Yen, reference librarian, Santa Barbara Public Library, Santa Barbara.

It wouldn't be possible to write a history book without other historians. I was two weeks away from submitting this manuscript when I learned that Bradley Agnew and Kelly Agnew had just released their book *John Gill Lemmon: Andersonville Survivor and California Botanist* about Sara Lemmon's husband. Not only is it an engaging read and a research tour de force, but they've been extraordinarily gracious about sharing information. I only wish I'd known about their work sooner.

Other historians and providers of welcome historical details include James Burns, Arizona Historical Society, Phoenix; Steven Carlson, Irvine, California; Bill Cavaliere, Cochise County Historical Society, Douglas, Arizona; Jan Cleere, Tucson; Bruce Dinges (retired), Arizona Historical Society, Tucson; Mike Eberhardt, Dallas; Bill Gillespie, retired U.S.

Forest Service archeologist, Tucson; Larry Ludwig (retired), Fort Bowie National Historic Site, Bowie, Arizona; Craig McEwan, Cochise County Historical Society; Mountain View Cemetery, Oakland; Deni J. Seymour, PhD, Tucson; Flint Speer, local historian, Shandon, California; and Jace Turner, community historian, Santa Barbara Public Library.

Although I'm an enthusiastic naturalist, I'm by no means a botanist—which means I learned a great deal from Sara and JG's work. The following people helped me understand it: Bob Behrstock, Hereford, Arizona; Dave Bertelsen, Tucson; Jack Dash, Desert Survivors, Tucson; Karen LeMay, Pollinator Corridors Southwest, Hereford; Elaine Moisan, Portal, Arizona; and David Stith, Phoenix.

For help in the technical areas, I'm grateful to art and paper conservator Susan Filter, Berkeley, for her knowledge and time assessing Sara's surviving artwork; Brendan Duddridge, Calgary, Alberta, Canada, for his patience and for allowing me to be a beta tester for Tapforms database while developing a catalog of Sara's paintings; and Gene Thomas, Vallejo, California, for his kind and valuable advice as I was preparing to photograph Sara's paintings.

Thanks to the generosity of various individuals and organizations, I was fortunate to be able to get some funding during this seven-year project: Arizona Native Plants Society (Cochise chapter, Tucson chapter); Bea Brooks Foundation for Creative Women; Border Community Alliance; Botanical Artists Guild of Southern California; Melanie Campbell-Carter; Cochise County Historical Society; Betty Creath; Dry River Poets; Money for Women (grant); Pat Parran; Mike and Cecil Williams; Mike Van Buskirk; and Winterhaven Village Townhouses, Tucson.

During the writing of this book I was able to take two nonfiction writing workshops. At the 2019 Tucson Festival of Books Masters Workshop, I was especially fortunate to share Sara's "shero's journey" with a gifted group of fellow students under the guidance of Stephanie Land. I also benefited from *Creative Nonfiction*'s online Historical Narrative class, led by Marty Levine, who provided valuable feedback.

And speaking of feedback, my fervent thanks go to the language-botany-and-art-loving friends, acquaintances, fellow writers, and colleagues who've offered encouragement, connections to editors and agents, ideas, homegrown produce, hospitality, suggestions, meals, coffee, access to resources, and advice. Some have even been willing—for years!—to read proposal after proposal, chapter after chapter. These generous souls include Jeff Babson, Daniel Brockert, Judie Bronstein, David Brooks, James Burns, Melanie Campbell-Carter, Francisco (Paco) Cantú, Rita Cantu, Daniel Connolly, Jillian Cowles, Alison Hawthorne Deming, Gail Hovey, Victoria Johnson, Charlie McKee, Nora Miller, Sharon Miller, Susan Cummins Miller, Alanna Mitchell, Carolyn Niethammer, Keith Pascoe (manager, Jack Ranch, Cholame, California), Mary Price, Deb Shaw, Jennifer Shopland, Carol Simon, Wendy Sizer, Nathan Smith, Nora Stark, Sandy Szelag and all the Dry River Poets, Liz Trupin-Pulli (JET Literary Associates, Santa Fe, New Mexico), Mike Van Buskirk, and Nickolas Waser.

To the nearly eight hundred recipients of the Sara Lemmon Project newsletter: Thank you for all your individual responses and for being my extensive and safely distanced advisory council.

I'm particularly grateful to Amy St. John, Sara's great-great-grandniece—a very special thanks to you for letting me rummage around in the attic of your family's history.

My heartfelt appreciation also goes to eagle-eyed copy editor Wayne Larsen for catching a significant number of mortifying auto-correct and author-induced errors. Any remaining mistakes or omissions are all mine.

I'm also indebted to current and former editorial and marketing staff members at Bison Books and the University of Nebraska Press, including Jackson Adams, Alicia Christensen, Tish Fobben, Haley Mendlik, Rosemary Sekora, and Emily Wendell. Designer Annie Shahan and formatter Mikala Kolander turned the cover and interior text into a truly beautiful book—many thanks for your work! Most of all I appreciate Bridget Barry, editor in chief—thank you for seeing the story that Sara could be and for helping it grow into something even better.

Selected Bibliography

1. At the Jumping Off Place

Bowman, J. N. "Driving the Last Spike: At Promontory, 1869." *California Historical Society Quarterly* 36, no. 2 (June 1957): 96–106, and 36, no. 3 (Sept. 1957): 263–74. Accessed March 5, 2020. http://cprr.org/Museum/Bowman_Last_Spike_CHS.html.

Lemmon, Sara Plummer. Letters and ephemera: Martha Everett St. John and Harold St. John Collection of Sara Plummer and John Gill Lemmon material. University and Jepson Herbaria, University of California, Berkeley. Accessed and photographed Jan. 2015, Jan. 2016, Jan. 2017. *Every chapter uses material from this source.*

St. John, Martha Everett. "Sara Allen Plummer." Personal obituary, n.d. Letters and ephemera: Martha Everett St. John and Harold St. John Collection of Sara Plummer and John Gill Lemmon material. University and Jepson Herbaria, University of California, Berkeley.

2. Perhaps You've Heard

Chronology of Santa Barbara. City of Santa Barbara. Accessed July 23, 2019. https://civicaweb.santabarbaraca.gov/services/community/historic/chronology.asp.

Haskell, Llewellyn. Correspondence with Sara Plummer. Letters and ephemera: Martha Everett St. John and Harold St. John Collection of Sara Plummer and John Gill Lemmon material. University and Jepson Herbaria, University of California, Berkeley. Accessed and photographed Jan. 2015, Jan. 2016, Jan. 2017.

Haskell Family Tree. Haskell Family History. Accessed March 29, 2020. http://www.haskellfamilyhistory.com/haskell/2/55188.html.

Kynett, Harold Havelock. "The Late Heat-Wave in New York." *Medical and Surgical Reporter* 75 (July–Dec. 1896): 428–29.

Loeffelbein, Robert. "Dio Lewis: He Started American Physical Education." *Improving College and University Teaching* 14, no. 1 (1966): 41–42. Accessed March 30, 2020. https://doi.org/10.1080/00193089.1966.10532496.

Plummer-Lemmon, Sara A. "Santa Barbara's First Library Efforts and Other Historical Sketches." *Santa Barbara Weekly Press*, March 10, 1910. Accessed Aug. 23, 2019. https://cdnc.ucr.edu/cgi-bin/cdnc?a=d&d=SBWP19100310 .2.50&e=-------en--20--1--txt-txIN--------1.

3. Like Death to Me to Be Idle

Plummer-Lemmon, Sara A. "Santa Barbara's First Library Efforts and Other Historical Sketches." *Santa Barbara Weekly Press*, March 10, 1910. Accessed Aug. 23, 2019. https://cdnc.ucr.edu/cgi-bin/cdnc?a=d&d=SBWP19100310 .2.50&e=-------en--20--1--txt-txIN--------1.

Wood, E. N. *Guidebook to Santa Barbara, Town and County, Containing Information on Matters of Interest to Tourists, New Settlers, Invalids, Etc.* Santa Barbara: Wood & Sefton, Book & Job Printers, 1872.

4. Botanist from the Sierras

Agnew, Brad, and Kelly Agnew. *John Gill Lemmon: Andersonville Survivor and California Botanist.* San Bernardino: Kindle Direct Publishing, 2020.

Bonta, Marcia Myers. *Women in the Field.* College Station: Texas A&M University Press, 1991.

Brewer, W. H., Sereno Watson, Asa Gray, and Geological Survey of California. *Botany.* Boston: Little, Brown, and Company, 1880. Accessed March 2020. https://babel.hathitrust.org/cgi/pt?id=ncs1.ark:/13960/t6d221s5r&view =1up&seq=582.

"The Civil War: Battle Unit Details; Union Michigan Volunteers (4th Regiment, Michigan Cavalry)." National Park Service. Accessed March 8, 2020. https://www.nps.gov/civilwar/search-battle-units-detail.htm ?battleUnitCode=UMI0004RC.

Copeland, Herbert F. "A Portrait of John Gill Lemmon." *Madroño* 5, no. 2 (1939): 77. Accessed March 31, 2020. http://www.jstor.org/stable /41422306.

Dupree, Hunter A. *Asa Gray: American Botanist, Friend of Darwin.* Baltimore: Johns Hopkins University Press, 1959.

Ertter, Barbara. "The Flowering of Natural History Institutions in California." Supplement 1, *Proceedings of the California Academy of Sciences* 55 (2004):

58–87. Accessed Nov. 28, 2020. http://www.norcalbotanists.org/files/B
  _ErtterLR.pdf.

Flint, Austin, ed. *Contributions Relating to the Causation and Prevention of Disease, and to Camp Diseases; Together with a Report of the Diseases, Etc. among
  the Prisoners at Andersonville, Ga.* Vol. 1 of *Sanitary Memoirs of the War of
  the Rebellion, Collected and Published by the United States Sanitary Commission.* New York: Published for the U.S. Sanitary Commission by Hurd and
  Houghton, 1867.

Guinn, J. M. *History of the State of California and Biographical Record of the
  Sacramento Valley, California: An Historical Story of the State's Marvelous
  Growth from Its Earliest Settlements to the Present Time; Also Containing
  Biographies of Well-Known Citizens of the Past and Present.* Chicago: Chap-
  man, 1906.

Lemmon, J. G. "In Memoriam: Amila Hudson Lemmon, Brief Sketch
  of Her Life and Labors." Compiled for *Pacific Rural Press*, Dec. 1885.
  Accessed March 8, 2020. https://babel.hathitrust.org/cgi/pt?id=mdp
  .39015071150448&view=1up&seq=9.

———. "Notes by a Pioneer Botanist." *Muhlenbergia* 4 (April 14, 1908): 17–21.

Michigan, Adjutant-General's Dept., Turner, G. H. *Record of Service of
  Michigan Volunteers in the Civil War, 1861–1865.* Kalamazoo MI: Ihling Bros.
  & Everard. Accessed March 8, 2020. https://babel.hathitrust.org/cgi/pt
  /search?q1=Lemmon;id=mdp.39015071160751;view=1up;seq=17;start=1
  ;sz=10;page=search;orient=0.

Rodgers, Andrew Denny, III. *American Botany, 1873–1892: Decades of Transition.* Princeton NJ: Princeton University Press, 1944.

Rudolph, Emanuel D. "Women in Nineteenth-Century American Botany:
  A Generally Unrecognized Constituency." *American Journal of Botany*
  69, no. 8 (Sept. 1982): 1346–55. Accessed March 8, 2020. https://bsapubs
  .onlinelibrary.wiley.com/doi/abs/10.1002/j.1537-2197.1982.tb13382.x.

5. My Dear, Soul-Knit Brother

California Academy of Sciences. Meeting minutes for Nov. 19, 1879. Rebekah
  Kim, personal communication, Feb. 28, 2020.

Carey, Andrew W. "Questions of Sovereignty: Pyramid Lake and the Northern
  Paiute Struggle for Water and Rights." PhD diss., University of New Mexico,
  2016. Accessed March 4, 2020. https://digitalrepository.unm.edu/cgi
  /viewcontent.cgi?referer=https://www.google.com/&httpsredir=1&article
  =1078&context=anth_etds.

Gray, Asa. Correspondence files of the Gray Herbarium. Correspondence with John Gill Lemmon. Botany, Libraries, Archives of the Gray Herbarium, Harvard University Herbaria, Cambridge MA. https://kiki.huh.harvard.edu /databases/botanist_index.html.

6. Into the Matrimonial Vortex!

Beidleman, Richard G. *California's Frontier Naturalists*. Berkeley: University of California Press, 2006.

Lemmon, J. G. Letter to John Muir. John Muir Correspondence. Online Archives of California. Accessed March 9, 2020. http://www.oac.cdlib.org /ark:/13030/kt4199r790/?order=2&brand=oac4.

"Marriage of Distinguished Scientists." *Santa Barbara Weekly Press*, Dec. 4, 1881.

7. The Heart of Santa Catalina

Bertelsen, C. David. *Thirty-Seven Years on a Mountain Trail: Vascular Flora and Flowering Phenology of the Finger Rock Canyon Watershed, Santa Catalina Mountains, Arizona*. Tucson: Desert Plants, 2018.

Bezy, J. V. *A Guide to the Geology of the Santa Catalina Mountains, Arizona: The Geology and Life Zones of a Madrean Sky Island*. Down-to-Earth, no. 83. Arizona Geological Survey, 2016.

Bowers, Janice Emily. *The Mountains Next Door*. Tucson: University of Arizona Press, 1991.

Brusca, Richard C., and Wendy Moore. *A Natural History of the Santa Catalina Mountains, Arizona (with an Introduction to the Madrean Sky Islands)*. Tucson: Arizona-Sonora Desert Museum Press, 2013.

Carroll, John Alexander, ed. *Pioneering in Arizona: The Reminiscences of Emerson Oliver Stratton and Edith Stratton Kitt*. Tucson: Arizona Pioneer's Historical Society, 1964.

Coulter, John M., and M. S. Coulter, eds. Editorial: "A Botanist's Marriage." *Botanical Gazette* 6, no. 1 (1880–81): 156. Accessed Dec. 13, 2019. https:// biodiversitylibrary.org/page/5173583.

Description of botanical research by Prof. J. G. Lemmon and wife. *Arizona Weekly Star*, May 12, 1881, pp. 3–4. Transcript at the Arizona Historical Society, Tucson.

Elliott, Wallace. *History of Arizona Territory: Showing Its Resources and Advantages; With Illustrations Descriptive of Its Scenery, Residences, Farms, Mines, Mills, Hotels, Business Houses, Schools, Churches, Etc. from Original Drawings*.

San Francisco: Wallace W. Elliott & Co., 1884. Transcript at Arizona Historical Society, Tucson.

Ewan, Joseph. "Bibliographical Miscellany—V. Sara Allen Plummer Lemmon
and Her 'Ferns of the Pacific Coast.'" *American Midland Naturalist* 32, no. 2
(1944): 513–18. Accessed Feb. 21, 2020. https://doi.org/10.2307/2421316.

Lemmon, J. G. "A Botanical Wedding Trip." *The Californian, a Western Monthly
Magazine* 4 (July–Dec. 1881): 512.

————. "Botanizing in Arizona." *Mining and Scientific Press: An Illustrated
Journal of Mining, Popular Science, and General News* 42, no. 25 (June
18, 1881): 397. https://www.google.com/books/edition/Mining_and
_Scientific_Press/EDe4PT1bmMQC?hl=en&gbpv=1&dq=JG+Lemmon
+golden+poppy&pg=PA397&printsec=frontcover.

8. An Extreme Outpost

Bennett, Peter S., R. Roy Johnson, and Michael R. Kunzmann. *An Annotated
List of Vascular Plants of the Chiricahua Mountains, Including the Pedregosa
Mountains, Swisshelm Mountains, Chiricahua National Monument, and Fort
Bowie National Historic Site.* Special Report no. 12. Tucson: U.S. Geological
Survey, Oct. 1996.

Jacobs, James Q. "Post-Contact of Social Organization of Three Apache
Tribes." Southwest Anthropology and Archeology Pages, 1999. Accessed
Feb. 16, 2021. http://www.jqjacobs.net/southwest/apache.html.

Kohrs, Donald G. "John Gill Lemmon and Sarah [*sic*] Allen Plummer Lemmon." In *Chautauqua: The Nature Study Movement in Pacific Grove, California*, 128–29. Pacific Grove CA: Donald G. Kohrs, 2015. Seaside: History of
Marine Science in Southern Monterey Bay. Accessed Feb. 16, 2021. https://
web.stanford.edu/group/seaside/pdf/ch5.pdf.

Lemmon, J. G. "Adventures in Apacheland." Draft article in Lemmon, Sara
Plummer, letters and ephemera, Martha Everett St. John and Harold St.
John Collection of Sara Plummer and John Gill Lemmon material. Univer-
sity and Jepson Herbaria, University of California, Berkeley.

Sweeney, Edwin R. *From Cochise to Geronimo: The Chiricahua Apaches, 1874–
1886.* Vol. 268 in The Civilization of the American Indian Series. Norman:
University of Oklahoma, 2012.

9. Eleven Days of Dungeon Life

Lemmon, J. G. "Perils and Pleasures of Botanizing in Arizona." In Lemmon,
Sara Plummer, letters and ephemera, Martha Everett St. John and Harold

St. John Collection of Sara Plummer and John Gill Lemmon material. University and Jepson Herbaria, University of California, Berkeley.

Rak, Mary Kidder. "The Hermit of the Chiricahuas." *Arizona Quarterly* 1, no. 2 (Summer 1945): 38–42.

Torrans, Thomas. "Southwestern History in the 'Arizona Quarterly,' 1945–1958: An Annotation on Contents." *Arizona and the West* 1, no. 3 (1959): 271–80. Accessed March 31, 2020. http://www.jstor.org/stable/40166966.

10. Happy in Our Work

Badè, William Frederic, ed. *The Cruise of the Corwin: Journal of the Arctic Expedition of 1881 in Search of De Long and the Jeannette by John Muir.* Boston: Houghton Mifflin Company; Cambridge: Riverside Press, 1917.

Lemmon, J. G. "Hints to Botanical Collectors." *Pacific Rural Press*, no. 8 (Feb. 1881).

11. Rushing, Reckless Life

"The Founding of Fort Huachuca." *Huachuca Illustrated: A Magazine of the Fort Huachuca Museum* 6 (1999): 1–186.

"Hesperocallis undulata A. Gray." USDA Natural Resources Conservation Service. Accessed March 16, 2020. https://plants.usda.gov/core/profile?symbol=heun2.

Parkhurst, Janet, R. Brooks Jeffery, Robin Pinto, Nancy Mahaney, Melissa Rees, Goran Radovanovich, Patricia Rogers, and Kris Jenkins (Preservation Studies class [ARC 4/597j]). "Historic and Architectural Resources of Downtown Tucson Arizona." College of Architecture, Planning and Landscape Architecture, The University of Arizona. Statement of Historic Contexts, National Register of Historic Places Continuation Sheet, U.S. Dept. of the Interior, National Park Service, 2002. Accessed Sept. 13, 2020. http://capla.arizona.edu/sites/default/files/projects/Historic%20and %20Architectural%20Resources%20of%20Downtown%20Tucson %20Arizona%20Context%20Study.pdf.

12. A Botanical Paradise

Bahre, Conrad J. "Land-Use History of the Research Ranch, Elgin, Arizona." Supplement 2, *Journal of the Arizona Academy of Science* 12 (Aug. 1977).

Lemmon, J. G. "Notes from Arizona." *Gardener's Monthly and Horticulturist* 24 (Sept. 1882): 276–77. Accessed Dec. 16, 2018. Google Books.

———. "Woodsia plummerae, n. sp.," in "Some Additions to the North American Flora." *Botanical Gazette* 7, no. 1 (Jan. 1882): 6–7. Accessed Dec. 16, 2018. https://archive.org/details/botanicalgazette78hano/page/n19.

Matheny, Robert Lavesco. "The History of Lumbering in Arizona before World War II." PhD diss., University of Arizona, 1975. https://repository.arizona.edu/handle/10150/565344.

Watson, Sereno. "Contributions to American Botany." In *Proceedings of the American Academy of Arts and Sciences,* 96–196. Boston: Metcalf and Co., 1882–83. Accessed Dec. 8, 2018. https://archive.org/details/proceedingsofame18amer/page/96.

13. A Fine Season's Work
Lowe, Edward Joseph. *Ferns: British and Exotic.* London: Groombridge and Sons, 1761. Accessed Feb. 17, 2021. https://archive.org/details/bub_gb_GUIAAAAAQAAJ/page/n3/mode/2up.

14. Lives Cast in Pleasant Places
Lemmon, J. G. "The Grand Canyon of the Colorado." *Overland Monthly* 12 (Sept. 1888): 244–56. Accessed March 19, 2020. http://quod.lib.umich.edu/m/moajrnl/ahj1472.2-12.069/278.

Munz, Philip, and Ivan M. Johnson. "Penstemon plummerae." *Bulletin of the Torrey Botanical Club* 49, no. 2 (Feb. 1922): 43. Accessed April 1, 2020. Google Books.

Stratton, R. B. *Captivity of the Oatman Girls: Being an Interesting Narrative of Life among the Apache and Mohave Indians.* 3rd ed. New York: Carlton & Porter, 1859. Accessed Aug. 28, 2020. https://archive.org/stream/capoatman00strarich#page/22/mode/2up.

15. Grandest Display
Andrews, A. *Report of A. Andrews, United States Commissioner for California at the World's Industrial and Cotton Centennial Exposition, New Orleans, Louisiana, December 16, 1884, to June 1, 1885.* Sacramento: State Office, 1886. Accessed Feb. 17, 2021. https://books.google.com/books/about/Report_of_A_Andrews_United_States_Commis.html?id=w0E9AQAAIAAJ.

Smith, Emory Evans. *The Golden Poppy.* San Francisco: Murdoch Press, 1901. https://books.google.com/books?id=ISIuAAAAYAAJ&dq=Lemmon+San+Luis+obispo+1887&source=gbs_navlinks_s.

16. Our Hillock in Cholame

Greene, Edward L. "Some Additions to Our State Flora." *West-American Scientist* 3, no. 28 (Aug. 1887): 157. Accessed March 15, 2020. Google Books.

Lemmon, John G. and Amabilis. Correspondence, 1887 to 1904. Clara Barton Papers: General Correspondence, 1838–1912. Manuscript Division, Library of Congress. Accessed Feb. 12, 2020. https://www.loc.gov/item/mss119730322/.

17. Life, to Me, Seems Sweeter

Kohrs, Donald G. *Chautauqua: The Nature Study Movement in Pacific Grove, California*. Pacific Grove CA: Donald G. Kohrs, 2015. Seaside: History of Marine Science in Southern Monterey Bay. Accessed Feb. 16, 2021. https://seaside.stanford.edu/Chautauqua.

Lemmon, J. G. *Pines of the Pacific Slope, Particularly Those of California*. In *Appendix to the Journals of the Senate and Assembly of the Legislature of the State of California*, 5: 68–150. Sacramento: State Office, 1889. Digitized Sept. 25, 2007. Accessed Feb. 17, 2021. Google Books.

Lemmon, John G., and Amabilis. Correspondence, 1887 to 1904. Clara Barton Papers: General Correspondence, 1838–1912. Manuscript Division, Library of Congress. Accessed Feb. 12, 2020. https://www.loc.gov/item/mss119730322/.

Watamull Foundation, Oral History Project. Interview with Dr. Harold St. John, Dec. 6, 1985. Honolulu HI: Watamull Foundation, 1987. Accessed Dec. 14, 2019. https://evols.library.manoa.hawaii.edu/bitstream/10524/48690/cropped_ocr_watumullohp_StJohn_combined.pdf.

18. The Narrowest Escape

California Federation of Women's Clubs. *Club Women of California: Official Directory and Register, Giving the Officers with Names and Addresses of All Members, Pub. under the Direct Supervision of the California Federation of Women's Clubs*. San Francisco, 1906–7. Google Books.

Ewan, Joseph. "Bibliographical Miscellany—V. Sara Allen Plummer Lemmon and Her Ferns of the Pacific Coast." *American Midland Naturalist* 32, no. 2 (1944): 513–18. Accessed Feb. 21, 2020. https://doi.org/10.2307/2421316.

A Friend of the School. "Training School in the General Hospital of San Francisco." *Trained Nurse and Hospital Review* 21, no. 1 (1898): 27–30. Accessed Dec. 27, 2018. Google Books.

Greene, Edward Lee. "New or Noteworthy Species, VII. *Calochortus plummerae.*" *Pittonia: A Series of Botanical Papers* 2 (July–Sept. 1891): 70. https://biodiversitylibrary.org/page/15255064.

Lemmon, John Gill. *Cone-Bearing Trees of the Pacific Slope North of Mexico and West of the Rocky Mountains. Hand-Book of West-American Cone-Bearers.* Oakland: Pacific Press Publishing Company, 1892. Author's copy contains handwritten note by Lemmon: "Corrected as it will appear in next edition and in larger volume. JGL."

"Lemmon, John Gill." In *Americanized Encyclopedia Britannica.* Chicago: Belford-Clarke Company, 1890, 10:6644. Accessed Jan. 7, 2020. Google Books.

Mabie, Adelaide. *Training Schools for Nurses in the State of California.* Vol. 1. San Francisco: Whitaker & Ray Company, 1899. Google Books.

"Mrs. Lemmon Vindicated." *San Francisco Chronicle,* March 21, 1889.

Stanton, Elizabeth Cady, Susan B. Anthony, Matilda Gage, Harriot Stanton Blatch, and Ida H. Harper. *The Complete History of the Suffragette Movement—(All 6 Books in One Edition): The Battle for the Equal Rights: 1848–1922 (Including Letters, Newspaper Articles, Conference Reports, Speeches, Court Transcripts & Decisions).* Accessed March 22, 2020. Google Books.

19. Sell Everything and Move

Hopkins, Louisa Parsons Stone. "Woman in Science." In *Art and Handicraft in the Woman's Building of the World's Columbian Exposition, Chicago, 1893,* edited by Maud Howe Elliott, 107–18. Chicago: Rand, McNally & Company, 1894.

James, George Wharton. *In and around the Grand Canyon: The Grand Canyon of the Colorado River in Arizona.* Boston: Little, Brown, and Company, 1900. Accessed March 24, 2020. Google Books.

20. A Sweet, Sacred Togetherness

Beidleman, Richard. "Lemmons and Poppies." *Jepson Globe* 14, no. 1 (Jan. 2004): 1.

George, Mary W., and Anna C. Murphy, under the direction of the State Board of Education. *Revised Third Reader.* Sacramento: State Printing Office. 1895.

Johnson, Eric Michael. "How John Muir's Brand of Conservation Led to the Decline of Yosemite." *Scientific American,* Aug. 13, 2014. Accessed Sept. 3, 2020. https://blogs.scientificamerican.com/primate-diaries/how-john-muir-s-brand-of-conservation-led-to-the-decline-of-yosemite/.

Lemmon, John Gill. "Conifers of the Pacific Slope—How to Distinguish Them." *Sierra Club Bulletin* 2:61–78; 156–73.

———. *How to Tell the Trees and Forest Endowment of Pacific Slope ... and Also Some Elements of Forestry with Suggestions by Mrs. Lemmon*. 1st ser. *The Cone-Bearers*. Oakland, 1902. http://hdl.handle.net/2027/uc2.ark:/13960/t0bv7c936.

21. Wish We Were Out in the Wild

"Botany Their Hobby, a Scientific Husband and Wife: Researches of Mr. and Mrs. Lemmon." *San Francisco Chronicle*, April 22, 1894.

Daniel, Thomas F. "One Hundred Fifty Years of Botany at the California Academy of Sciences (1853–2003)." *Proceedings of the California Academy of Sciences*, ser. 4, vol. 59, no. 7 (2008): 215–305.

"The Golden Poppy Is Our Emblem." *Oakland Enquirer*, March 3, 1903. Clipping, John Gill Lemmon and Sara Plummer Lemmon Papers, University and Jepson Herbaria Archives, University of California, Berkeley.

Lemmon, Sara A. P. *A Record of the Red Cross Work on the Pacific Slope: Including California, Nevada, Oregon, Washington, and Idaho with Their Auxiliaries; Also Reports from Nebraska, Tennessee and Far-Away Japan*. Oakland: Pacific Press Publishing Co., 1902.

"Press Reception to Honor General Shafter." *San Francisco Call*, Nov. 19, 1904.

Smith, Emory Evans. *The Golden Poppy*. San Francisco: Murdoch Press, 1901. Google Books.

"The State Flower of California." *Western Journal of Education* 8 (March 1903):130. Accessed Feb. 1, 2020. Google Books.

Thompson, Erwin N. *Defender of the Gate: The Presidio of San Francisco; A History from 1846–1995*. Historic Resource Study. Golden Gate National Recreation Area CA, National Park Service, July 1997.

22. Safe—Tho' Tremendously Shaken

"Came to Tucson on the First Railroad Train: Celebrated California Botanist and Wife Celebrating Silver Wedding Anniversary Where Honeymoon Was Spent." *Tucson Citizen*, June 30, 1905. Accessed Jan. 29, 2020. https://www.newspapers.com/image/580146133/?terms=Lemmon.

Crosswhite, Frank S. "'J. G. Lemmon & Wife,' Plant Explorers in Arizona, California, and Nevada." *Desert Plants* 1, no. 1 (Aug. 1979): 12–21.

Debaksey, Dale. "A Bay of Botany: Alice Eastwood's Nine Decades and Three Hundred Thousand Specimens." *Women You Should Know* (blog), Sept. 19, 2018. Accessed Feb. 6, 2020. https://womenyoushouldknow.net/botany -alice-eastwood/.

"Eats Luncheon on Continent's Edge: President Pays Visit to the Cliff House and Shows That His Appetite Is Sharpened by California's Pure Air." *San Francisco Chronicle*, May 14, 1903.

Lemmon, J. G. "Forest Endowment of Pacific Slope." *Out West: A Magazine of the Old Pacific and the New* 24 (March 1906): 172–99. Accessed Feb. 4, 2020. Google Books.

Li, Fay-Wie, Kathleen M. Pryer, and Michael D. Windham. "Gaga, a New Fern Genus Segregated from Cheilanthes (Pteridaceae)." *Systemic Botany* 37, no. 4 (Oct.–Dec. 2012): 845–60. Accessed Feb. 3, 2020. https://www.jstor.org /stable/23362703?read-now=1&seq=11#page_scan_tab_contents.

"Old Times Are Recalled: Prof. and Mrs. Lemmon Visit Santa Barbara after Many Years' Absence." Clipping from "Santa Barbara paper," June 8, 1905. University and Jepson Herbaria Archives, University of California, Berkeley.

"Ruth Newman, a Survivor of the 1906 San Francisco Earthquake, Dies at 113." *New York Times*, Sept. 2, 2015. Accessed Feb. 6, 2020. https://www.nytimes .com/2015/09/03/us/ruth-newman-a-survivor-of-the-1906-san-francisco -earthquake-dies-at-113.html.

Thomas, Gordon, and Max Morgan-Witts. *The San Francisco Earthquake: A Minute-by-Minute Account of the 1906 Disaster*. Open Road Integrated Media, 2014. First published by Stein and Day, 1971.

Wilson, Carol Green. *Alice Eastwood's Wonderland: The Adventures of a Botanist*. San Francisco: California Academy of Sciences, 1955.

23. I Feel So Helpless and Alone

Kaufman, Polly Welts. *National Parks and the Woman's Voice: A History*. Updated ed. Albuquerque: University of New Mexico Press, 2006. Accessed Aug. 23, 2019. Google Books.

Kibbe, Alice L. *Afield with Plant Lovers and Collectors*. Carthage IL: printed by the author, 1953.

Plummer-Lemmon, Sara A. "Santa Barbara's First Library Efforts and Other Historical Sketches." *Santa Barbara Weekly Press*, March 10, 1910. Accessed

Aug. 23, 2019. https://cdnc.ucr.edu/cgi-bin/cdnc?a=d&d=SBWP19100310
.2.50&e=-------en--20--1--txt-txIN--------1.

"Professor Lemmon Pens Book on Trees: Will Describe All Timber in West
from Alaska to Mexican Line." *San Francisco Call,* July 7, 1907.

24. Partners in Botany

"Death of Mrs. J. G. Lemmon." *Pacific Unitarian* 32 (Feb. 1923): 40. Accessed
Feb. 1, 2020. Google Books.